India's Climate Change Identity

Samir Saran • Aled Jones

India's Climate Change Identity

Between Reality and Perception

Samir Saran
Observer Research Foundation
New Delhi, India

Aled Jones
Global Sustainability Institute
Anglia Ruskin University
Cambridge, United Kingdom

ISBN 978-3-319-46414-5 ISBN 978-3-319-46415-2 (eBook)
DOI 10.1007/978-3-319-46415-2

Library of Congress Control Number: 2016951947

© The Editor(s) (if applicable) and The Author(s) 2017
This work is subject to copyright. All rights are solely and exclusively licensed by the Publisher, whether the whole or part of the material is concerned, specifically the rights of translation, reprinting, reuse of illustrations, recitation, broadcasting, reproduction on microfilms or in any other physical way, and transmission or information storage and retrieval, electronic adaptation, computer software, or by similar or dissimilar methodology now known or hereafter developed.
The use of general descriptive names, registered names, trademarks, service marks, etc. in this publication does not imply, even in the absence of a specific statement, that such names are exempt from the relevant protective laws and regulations and therefore free for general use.
The publisher, the authors and the editors are safe to assume that the advice and information in this book are believed to be true and accurate at the date of publication. Neither the publisher nor the authors or the editors give a warranty, express or implied, with respect to the material contained herein or for any errors or omissions that may have been made.

Cover illustration: Détail de la Tour Eiffel © nemesis2207/Fotolia.co.uk

Printed on acid-free paper

This Palgrave Macmillan imprint is published by Springer Nature
The registered company is Springer International Publishing AG
The registered company address is: Gewerbestrasse 11, 6330 Cham, Switzerland

Preface

Climate negotiations and policy have been shaped by many individuals representing various interest groups over the past 30 years. Whether these interest groups are nation states, businesses, peoples or charities their approach to engagement and constructive or obfuscatory dialogue shapes the outcome.

Many argue that the vested interests of these different groups are fundamentally incompatible while others hold out for inspirational political leadership that can both compromise and drive ambition. However, some of the represented interests go much deeper than others and occasionally one set of individuals can hold a plethora of competing interests that are not always mutually reinforcing.

One such group are those individuals who represent one of the largest nation states in the world—India. India's interests are complex and ever changing. Indeed, its interests are more than just 'interests'—they represent a set of identities that shape India itself. Identities are more difficult to compromise on or negotiate away and are therefore vital to understand whether you are an academic, negotiator or citizen.

This book scans India's development landscape to decipher its key developmental identities, namely *rural identity, energy security identity, industrial identity, entrepreneurial identity, developing nation identity* and *emerging nation identity*. This basket of identities coexists alongside a diverse range of motivations, perceptions and expectations that influence India's domestic development agenda and the global emissions trajectory. Importantly, it challenges any attempt to construct a cogent Indian position at international climate change negotiations. We attempt to uncover and critically analyse the role and presence of these developmental identities within the climate change discourse in India, and their influence on India's negotiating stance at international climate forums. This allows us to evaluate the central hypothesis, which argues that India struggles to tackle its energy requirements, larger development goals and climate change imperatives, all together, due to the variety of factors it needs to accommodate while formulating its policies and negotiating stance. We locate the Indian identity on climate change at the proverbial golden median of a fluid triangle evolving over time. The triangle is broadly shaped by three unmistakable vectors: India's aspiration for global leadership; its growing appetite to create an industrial and entrepreneurial economy; and the continuing attempt to respond to the challenges of mass poverty and deprivation.

This identity-based approach to understand the motivations and drivers of a nation's response to climate change imperatives is a humble but significant contribution to the study of the politics of climate change. We outline this approach in Part I of the book, which covers (a) the background and review of India's development narrative and its increasing interplay with climate change conversations; (b) India's struggle for meeting the burgeoning energy demand and needs, in the backdrop of limited availability, capacity and distribution asymmetry; and (c) India's more recent business and policy emphasis on renewable energy and energy efficiency and its impact on the climate discourse. This leads to the construction of six distinct identities.

Part II of the book is divided into three sections. In the first section, we present the findings from a media content analysis of the climate change discourse in India. The second section comprises interviews with senior Indian climate experts. The third section comprises analysis of the speeches made by the official Indian climate interlocutors. Together, they offer a holistic understanding of the outlook and attitude of various parties who shape India's response to climate change.

These are thereafter synthesised in a discussion and a conclusion in Part III.

Acknowledgements

This book is adapted from my doctoral research at the Global Sustainability Institute (GSI). I must therefore first thank the organisation that I belong to, the Observer Research Foundation, and particularly its Director, Sunjoy Joshi, for making that research endeavour possible. My colleagues supported me, motivated me and helped me reshape and refine my approach to this effort. Two of them—Vivan Sharan and Sonali Mittra—need special mention and a big thank you.

My one regret is that my father is no longer with me to witness the end of this journey, but I draw satisfaction from the fact that he knew I was on my way. This piece of work is a dedication to his romantic belief in the power of education and his investments in mine. A big thanks to my mother, who has constantly prodded and encouraged me as well as my brother Suvir and sister Seema, who have always taken pride in most things I do. A very special thanks to my partner and now wife, Katharina, who has been a huge pillar of strength and support, and was able to magically conjure up time in my calendar that allowed us to tie the knot over a big fat Indian wedding a fortnight before I defended my dissertation.

Finally, this book would not be possible without my co-author Aled Jones, who as my doctoral supervisor was politely demanding and cruelly clever and sent me back to the drawing board on a number of occasions. He was the reason I came to GSI and co-authoring this book with him marks the culmination of an intellectually rewarding journey.

Contents

Part I Identity and Climate Change

1 Ontology of the Self — 3

2 India and Climate Change — 15
Growth and Development Trajectory — 17
Development and International Politics — 22
Energy, Development and Growth — 27
  Energy Access and Affordability — 27
  Energy Production — 31
  Energy Security — 37
Delineating the Identities — 38

Part II Uncovering Indian Climate Identities

3 Media Coverage in India — 45
India's Climate Change Identity Before COP-15 in 2009 — 48
India's Climate Change Identity During COP-15 in 2009 — 49
India's Climate Change Identity After COP-15 in 2009 — 50
India's Climate Change Identity in 2009 — 51
India's Climate Change Identity Before COP-18 in 2012 — 53
India's Climate Change Identity During COP-18 in 2012 — 54

India's Climate Change Identity After COP-18 in 2012   55
India's Climate Change Identity in 2012   57
India's Climate Change Identity Across Five Mainstream Newspapers During COP-15, 2009   58
India's Climate Change Identity Across Five Mainstream Newspapers During COP-18, 2012   60
Auxiliary Findings of 2009 and 2012   63

4 The Personal Choices of Indian Experts   65
   Rural Identity   65
   Energy Security Identity   72
   Industrial Identity   76
   Entrepreneurial Identity   78
   Developing Nation Identity   81
   Emerging Nation Identity   85
   Trends in Expert Opinions   88
     Climate Colonialism   93
     New Opportunities   94
   Creating Leverage Within a Monolithic Identity   98

5 India's Official Stance   99
   Evolving Identities; Congruent Identities   103
   Discovering the 'I'   107

Part III   The Modi Factor and India's Future Identity

6 Looking Ahead   111
   New Delhi Is Not India   113
   The Modi Factor   116
   In Sum   122
     Coal Will Fuel Growth in the Foreseeable Future   125
     Enterprise Has to Complement It   125
     India Is Ready for a Seat at the Global High Table   126

Bibliography   129

Index   143

Acronyms

APDRP	Accelerated Power Development and Reforms Programme
ASSOCHAM	Associated Chambers of Commerce and Industry in India
AT&C	Aggregate Technical and Commercial
BASIC	Brazil, South Africa, India and China
BRICS	Brazil, Russia, India, China and South Africa
CEA	Central Electricity Authority
CAGR	compounded annual growth rate
CBDR	Common but Differentiated Responsibilities
CII	Confederation of Indian Industry
CIL	Coal India Limited
COP	Conference of Parties
DGCI&S	Directorate General of Commercial Intelligence and Statistics
G20	a group of 20 major economies represented by the governments and central bank governors, which coordinates international economic policies
G-77	Group 77 developing countries forming the largest intergovernmental organisation in the United Nations
GDP	gross domestic product
GHG	global greenhouse gas
GNI	gross national income
HDI	Human Development Index
ICR	intercoder reliability
IEA	International Energy Agency
IMF	International Monetary Fund
MCA	media content analysis
MoEFCC	Ministry of Environment, Forests and Climate Change
MoPNG	Ministry of Petroleum and Natural Gas

MPCE	monthly per capita expenditure
MSP	minimum support price
MSPI	Ministry of Statistics and Programme Implementation
MT	million tonnes
$MTCO_2E$	million tonnes of carbon dioxide equivalent
MTOE	million tonnes of oil equivalent
NAM	Non-Aligned Movement
NCMP	National Common Minimum Programme
NEP	National Electricity Policy
NSS	National Sample Survey
ODA	official development assistance
OECD	Organisation for Economic Co-operation and Development
PDS	Public Distribution System
PFC	Power Finance Corporation
Q & A	question and answer
R/P	reserves to production
RPO	Renewable Purchase Obligation
SEIAA	State Environment Impact Assessment Authority
UNCED	United Nations Conference on Environment and Development
UNCHD	United Nations Conference on Human Development
UNCHE	United Nations Conference on Human Environment
UNCTAD	United Nations Conference on Trade and Development
UNDP	United Nations Development Programme
UNFCCC	United Nations Framework Convention on Climate Change
WEO	World Energy Outlook
WTO	World Trade Organisation

List of Figures

Fig. 2.1	Key indicators of India's development	21
Fig. 2.2	Aggregate technical and commercial losses in the power utilities sector (an illustrative sample of states)	22
Fig. 3.1	MCA 2009: Before, during and after COP-15 reportage (percentage of all analysed media articles including each identity)	62
Fig. 3.2	MCA 2012: Before, during and after COP-18 reportage (percentage of all analysed media articles including each identity)	63
Fig. 4.1	Expert opinions (analysis of the interviews)	92
Fig. 4.2	Interview analysis: India's primary fuel demand and climate change	95
Fig. 4.3	India's entrepreneurial identity	96
Fig. 4.4	Interview analysis: India's rural identity	97
Fig. 4.5	Interview analysis: India's entrepreneurial identity	97
Fig. 5.1	Key themes from the Indian Official Speeches made at the Global Climate Change Forum	101

List of Tables

Table 2.1	India – Population over the years and projections	18
Table 2.2	India: Rural–urban distribution of population with household size and monthly per capita expenditure (MPCE)	19
Table 2.3	Expenditure classes in India (2009–10)	20
Table 2.4	India and the world – Access to electricity and population	28
Table 2.5	The scope for increasing end-use efficiency through demand side management in India	36
Table 3.1	Languages of the Indian newspapers (2013–2014)	46
Table 3.2	Media content analysis for reportage on climate change before COP-15 (before 7 December 2009)	48
Table 3.3	Media content analysis for reportage on climate change during COP-15 (7–18 December 2009)	49
Table 3.4	Media content analysis for reportage on climate change after COP-15 (18 December 2009 onwards)	50
Table 3.5	Media content analysis for 2009 reportage on climate change (COP-15 aggregate)	52
Table 3.6	Media content analysis for 2012 reportage on climate change before COP-18 (before 26 November 2012)	53
Table 3.7	Media content analysis for 2012 reportage on climate change during COP-18 (26 November–7 December 2012)	55
Table 3.8	Media content analysis for 2012 reportage on climate change after COP-18 (7 December 2012 onward)	56
Table 3.9	Media content analysis for 2012 reportage on climate change (Doha 2012)	57
Table 3.10	Media content analysis for five mainstream newspapers reporting on COP-15, 2009	58

Table 3.11	Media content analysis for five mainstream newspapers reporting on the COP-18, 2012	60
Table 3.12	Auxiliary findings of COP-15, 2009 and COP-18, 2012 MCA	64
Table 4.1	Does agriculture sector influence India's position at climate negotiations?	66
Table 4.2	Can agriculture subsidies be removed? (Yes or no and explain why)	68
Table 4.3	Is there scope for energy efficiency savings in the agricultural sector through demand side management?	70
Table 4.4	How will India respond to climate-induced vulnerabilities in the agriculture sector?	71
Table 4.5	The development of the renewable energy sector is inevitable? (Yes or no and explain why)	73
Table 4.6	Given that a large share of primary energy generation is based on coal (and India has among the world's largest reserves), how will India approach issues around coal production and consumption?	74
Table 4.7	Should hydro and nuclear energy be prioritised given extant resource constraints?	75
Table 4.8	India's growth over the last two decades has been driven by the service sector. However, for sustainable, inclusive growth, India must generate employment through development of industry (such as manufacturing). How will India manage this vital imperative?	77
Table 4.9	Is India positioning itself to be a large consumer and producer of green energy and clean technologies? (Yes or no and explain)	78
Table 4.10	Can rapid development of industrial efficiency be envisioned without strong policy interventions at the federal level?	79
Table 4.11	What are the drivers of industrial efficiency in the Indian economy?	80
Table 4.12	Will being a part of G-77 be beneficial to India in climate change negotiations over the next decade?	82
Table 4.13	How important is the South–South cooperation within the context of global leadership in climate change mitigation?	83
Table 4.14	India is a trillion-dollar economy and is significantly larger than most economies that are part of the G-77. Is this a contradiction in itself?	83
Table 4.15	Will the principles of common but differentiated responsibility and equity continue to be the cornerstones of Indian negotiating policy?	84

Table 4.16	Is it incumbent upon India to position itself as a leader in climate change negotiations or is it strategically more logical for it to manage its interest through closer coordination with other developing countries?	85
Table 4.17	How must India position itself at climate change negotiations?	86
Table 4.18	Will India continue to coordinate interests with China/BASIC within the context of negotiating a new climate change agreement applicable from 2020?	87
Table 4.19	Can/will India bind itself to legal commitments?	88
Table 5.1	Textual analysis findings	104

PART I

Identity and Climate Change

CHAPTER 1

Ontology of the Self

Abstract This chapter explores the concept of identity and the 'self'. In particular, it expands on the assessment of India through an identity lens. It goes on to highlight the key drivers for India that help shape and evolve its self-identification. We briefly outline some of the key milestones in the international climate negotiations to set the scene for the book.

Keywords Identity · Climate negotiations · UNFCCC · Growth

German mathematician and philosopher Gottfried Wilhelm von Leibniz defined 'perception' as the 'representation of a multitude in a unity' (Puryear 2008). The unique ability to perceive or to understand someone or something in the way that we do, by ascribing language, traits and characteristics (or a combination of properties), is what makes us human. It follows, therefore, that the ability to understand animals from humans is the ability to understand the philosophical 'other'. This human ability to aggregate multiple stimuli into a unity is perhaps something that is specifically useful for policy studies. This unity is distinct from singularity, as the multitude continues to coexist.

French philosopher and existentialist Emmanuel Levinas grappled with the notion of the 'other' and its intrinsic and unequivocal relation to the self. He remarked that the 'I is identification in the strong sense; it is the very phenomenon of identity' (Levinas 1963). His basis for preponderance on the phenomenon of identity in the development of the idea of the self derives from an ethic that orients the subject towards responding to the ethical, wherein the ethical itself is not simply characterised as a system of morals but by a prior sense that the 'other makes possible the transformation from subject to subjectivity' (Sarukkai 1997).

For Levinas, identity was labile and non-permanent. Identity derives not from some property with which the self identifies, but rather, the self is the same as the properties with which it identifies. That is, as Levinas delicately puts it, 'the outside of me solicits it in need: the outside of me is for me' (Levinas 1963). In everyday life, we come to understand the act of observation as one which is implicit in the process of perceiving. That is, perception is understood falsely to be, if Levinas' intellection is to be applied, a process in which the properties of the world are independent of the observer. Stated differently, our observations and our lived experience are indistinguishable.

Biologist Humberto R. Maturana, in his essay titled 'Ontology of Observing', has gone a step further and treated the phenomenon of cognition described earlier as a biological phenomenon, which takes cognition and language as 'phenomena of our human domain of experiences that arise in the praxis of our living' (Maturana 1988). Like Levinas, he asserts that human beings exist as self-conscious entities in a physical domain, which exists in turn only as secondary to the 'happening of living of the human observer' (Maturana 1988). The observer is the reflective axis of being, around which physical reality gets confected. For Maturana, the physical domain is an agglomeration of cognitive entities, including nature, religion and atoms, with cognitive entities themselves being only mere observations of happenings from the perspective of the human observer. His premise aligns with Levinas in that everything that happens is human responsibility, as it is simply the observation of happenings that defines the world around us.

Psychologists Rogers, Kell and McNeil's (1948) acclaimed study titled 'The Role of Self Understanding in the Prediction of Behaviour', which hypothesises the impact of factors influencing early identity creation to adjustment in later life, has deliberated at length on the significance of identity, Levinas' 'I'. Aside from a factor they label 'self-insight', the team tested factors, including social experiences, mentality, heredity, family environment, economic and cultural environment, and physical education and training. The study found a strong correlation between positive self-insight and better later adjustment. The authors found, to their own surprise, that 'self-insight' – an individual's understanding of self – played an extremely dominant role in the process of later life adjustment.

It is not often that two branches of social science reaffirm each other's findings. Much like Levinas, Rogers et al. (1948) posit that the idea of 'self' is the most important variable of perception. Levinas' theorising and Rogers' study converge on the premise that objectivity has no meaning without a frame of reference. The implication is clear – there is no objective truth, only what is perceived to be the objective truth by the observer, who relies primarily upon self to ascribe any value or property to the observation. The observed is not independent from the observer and the narratives that shape the self and its beliefs are indeed flowing from its understanding of itself, which in turn is based on the unitary assimilation of a multitude of lived experiences.

Maslow's theorisation of 'self-actualisation' offers insights into how the self is constituted. For Maslow, people cannot be studied in isolation, and are whole functions rather than functions of uniquely identifiable parts. As per Maslow, it is impossible to isolate parts of this melange to analyse people or their decisions through the study of independent strands. Maslow describes self-actualisation as 'the intrinsic growth of what is already in the organism, or more accurately, of what the organism is' (Maslow et al. 1970). Maslow's work in this domain begins to uncover the 'social aggregation' and group-think that create collective underpinnings and shape individual preferences and responses.

Self-identification, which delves into how different stakeholders within India respond to a specific set of stimuli, in turn influences the policy discourse. In this instance, the stimuli considered are the various strands of the global climate change discourse in the backdrop of India's specific development challenges. And the 'I' is aggregated into the 'we' by collecting similar impulses of self-identification.

This aggregation of individual preferences and perceptions into a collective is perhaps the enduring ethos of arranging societies, communities and even the politics of provinces and nations. If this is attempted on the issue of climate change, the process can be acrimonious as well. The process of collective identification on climate change, its impact and possible responses is contested within and across countries.

Not surprisingly, domestic policies are bitterly disputed in India, and common ground (at any level of aggregation) on assessment and action remains elusive. The intention is not to develop a grand unified Indian identity and approach on the subject of climate change, rather to collate different strands of perceptions and perceived objectivity, discussed in the previous paragraphs, that different groups (civil society, government, private sector, academia, etc.) relate to.

Most such groups are not homogenous either. For instance, within the government, different states (provinces or federal units) have significant divergences with the central government and within themselves. The resource-rich states of India have unique preferences. The industrialised states develop their awareness based on their lived experience and the agrarian groups respond to their existence. The same would be true for different private sector enterprises, academia and civil society entities.

The primary focus is on national-level conversations (influenced though they may be by other factors as well). The discussions presented are based on media analysis of national mainstream newspapers, studying policy pronouncements by key Indian interlocutors on climate change at international forums, and interviewing individuals across various interest and professional groups engaged in the climate policy discourse in the national capital, New Delhi.

Such a study allows for accommodating, understanding and making meaning of the multiple viewpoints from diverse stakeholders that contribute and inform to India's voice in the debate, or the 'representation of a multitude in a unity' as per Leibniz (Puryear 2008). It also helps decipher the observation–identity–calculus poignantly described by Anaïs Nin (1961), specifically for Indian (national) identification with climate change and the dominant Indian identity(ies), as it negotiates at international forums.

This is vital for global negotiations. India has a virtual veto on any climate action due to the size of its economy, the second largest population and increasing emissions aggregated at the national level. To negotiate and respond to Indian needs, positions and observations, it may be useful to deconstruct the lived experience that finally shapes observations, perceptions, international positions and action.

Carl Rogers and Abraham Maslow in the 1960s suggested how perception of self-shapes responses and behaviours (Bandura 1994). Self-efficacy and associated identities form a critical part of decision-making (McClelland 1985). Put differently, people's judgement of their capabilities and capacities to organise and execute courses of action is critical in attaining designated types of performances. Self-perception theories further explain attitude formation based on response–outcome expectations and desires. Behaviouralists draw the focus towards the role and influence of external environment (stimuli), differentiated into classical and operant conditioning in shaping responses (Ford and Urban 1963). Humanistic approach, however, is more holistic rather than reductionist, and better explains the impact of a person's consciousness on their own identity, which in turn shapes their actions and responses. Considered as a breakthrough in human psychology in terms of its holistic view of individuals, this theory, when applied to collective identity and responses, may be beneficial in bridging the knowledge gap for logic of negotiations in international climate change dialogues. For any sustainable climate agreement, it is important discovering the 'I' (identity) in India's climate change conversations and propositions.

It may be useful at this point to reflect briefly on India's evolving posture at climate negotiations. This can be examined by analysing the official statements of the Indian government at international climate forums and existing research and commentary on the subject.

In an incredibly complex climate change negotiation process and its evolution, India's approach has been in a manner predictable and contentious at the same time. At the United Nations Conference on the Human Environment (UNCHE) in 1972, Prime Minister Indira Gandhi articulated what would become India's enduring position, noting that the environment cannot be improved in conditions of poverty. This laid the foundation of India's subsequent position at the climate change negotiations, which demanded attention towards poverty eradication, common but differentiated responsibilities, environmental justice and equity, financial flows, technology transfer and mitigation assistance. It is argued by many that the poverty–income narrative of the Indian climate change discussion has retained its spotlight till date, despite the unprecedented growth that has occurred over the past two decades.

In 1972, at the UNCHE in Stockholm, India's prioritisation of poverty eradication and development aspirations was noticeable. By 1992, at the United Nations Conference on Environment and Development, Rio de Janeiro, rallying together popular support against the notion that developing countries share equivalent responsibility in mitigating climate change impacts, India strongly maintained the principle of common but differentiated responsibilities and equity.

While demanding autonomy in deciding the development pathway for itself, India clearly spelt out the funding and technology transfer obligations of the developed nations for mitigation of climate change effects. Presumably based on the divergent principles of per capita emission and equitable burden sharing, resource sharing remains a highly contested area in climate talks. India's energy security concerns also came to the forefront in the 1992 negotiations.

India's per capita emission is less than a third of the global average (World Bank 2013). Policymakers understand that India will need to meet its targets on poverty, unemployment and literacy, and provide electricity access to the 44% of its population currently doing without. This will require much greater energy use alongside the consequent increased emissions.

In 2010, the then Minister of State for Environment and Forests Jairam Ramesh in his address to parliament noted that 'as India's Gross Domestic Product (GDP) grows, its emissions shall increase in absolute terms but the growth rate of emissions may moderate as reflected in the declining intensity of GDP. In fact, India's emission intensity has declined by 17.6% between 1990 and 2005 while its energy intensity has been decreasing since the 1980s and is already in the same range as that of the least energy intensive countries in the world' (Ramesh 2010a). It can be argued that some of this is due to the structural constraints to access energy and partly by design.

The rhetoric of energy demands and inclusive growth imperatives is profoundly nestled in the global climate change negotiations. However, in 2009, India signed the Copenhagen Accord and volunteered to reduce the emission intensity of its GDP by 20–25% by 2020 in comparison to the 2005 levels (Ministry of Environment and Forest 2010). The then Prime Minister Dr. Manmohan Singh further asserted that India's per capita emission would never exceed that of the developed countries (Ministry of External Affairs 2007). This in many ways was a breach of the firewall that was built by India against owning up to any responsibility towards climate mitigation.

India has shown leadership in voicing developing nations' concerns constantly. At the Conference of Parties (COP-19) in Warsaw in 2013, Jayanthi Natarajan, India's then Minister for Environment and Forests, re-emphasised the need for a pro-poor development agenda and principles of equity (Natarajan 2011).

There is scepticism over India's role in climate change negotiations. Some allege that India is playing a defensive and largely reactionary role in international climate change negotiations, steadfastly refusing to countenance any form of international legal commitment to reduce its own global greenhouse gas emissions. On the other hand, it can be argued that since 2010, India's approach to international climate diplomacy has been shifting towards more constructive engagement with international partners and the United Nations Framework Convention on Climate Change (UNFCCC) (Hallding et al. 2011).

Irrespective of how the Indian position is analysed, and spliced, it is quite apparent that India's evolving self-identification is likely to be shaped by four critical drivers:

a) The primary driver will be the 'need', that is the political obligation and the social responsibility, to cater to the millions living in poverty, outside the economic mainstream, physically and sociologically removed from the assertive middle class and elite, who typically shape India's national response to the global discourse on climate change. This need-based discourse of political populism has been deliberated at length by Subramanian (2007), who postulates that populism will continue to dominate Indian realpolitik.

b) The second driver will be the urge for greater acceptability and influence on the global power stage, especially in the Global South, India's traditional domain of influence now under pressure due to Chinese overplay; including among its partners in the Non-Aligned Movement and Group of 77 countries; and among its new partners in the BRICS (Brazil, Russia, India, China, South Africa) and extended BRICS, such as Indonesia and Mexico. As per Stuenkel (2010), this could be traced to the evolution of India's worldview from an 'idealist, Nehruvian perspective' to a 'pragmatic realpolitik-driven attitude' (Stuenkel 2010).

c) The third driver will be the unique political space and mind space it seeks to occupy for itself at the global high table, which will derive legitimacy from its contra-distinction to the proverbial 'other', the Global North, comprising the OECD (Organisation for Economic Co-operation and Development) and rest of the developed world. While the OECD has started engaging with India closely, India is not a member, and neither are any of the other countries in the BRICS grouping (Saran and Sharan 2012a). The struggle to define itself in relation to the 'other' is certainly visible in varying degrees in relation to its developing country persona and that of an emerging power.

d) Finally, as the country transits towards an industrialised and developed society, the huge demand for additional resources created in its wake would coincide with the concurrent challenge to extend and build efficiencies in the supply of the very basic of public services.

The imperatives for growth and modernisation, essentialised by liberal and unencumbered environment and merit, will compete with the more basic impulse of promoting policies that are designed as tools for state patronage. It has been argued that if India's industrial sector does not grow, then modernisation and socio-economic progress could prove elusive, while the patronage system (based on selective handouts of industrial licences and permits) continues to impinge on India's huge competitive advantage of comparatively lower labour, which could be a game changer (Felipe et al. 2010).

The period between 2009 and 2012 was a very critical phase in India's involvement in the international climate change negotiations. At COP-15, held at Copenhagen in 2009, India, along with Brazil, Russia, and China played a pivotal role in collectively articulating a perspective that was consonant with India's historical positions at climate change negotiations since the early days of climate talks.

In fact, the early days of India's active involvement in global climate change negotiations was marked by consensus building with emerging nations, notably China, and to a lesser degree with other members of the BASIC grouping (Brazil, South Africa, India and China), on negotiating positions and outcomes. Arguably, the unconscious impulses informing the action and choice of partners even at the outset were aligned to the 'emerging power' and 'developing nation' self-identities (Ramesh 2009).

Several studies have examined India's engagement in the international climate change negotiations through the prism of identity, for example Hurrell and Sengupta (2012), Qi (2011) and Dubash (2013). However, these studies are largely framed around the emerging economy identity and the attendant power discourse. Dubash (2013) describes India's attempt at grappling with its two distinct yet concurrent realities in the multilateral sphere as an 'intriguing' duality of bargaining positions. He notes that while, on the one hand, India is a developing country with a large population of poor, with one of the lowest per capita emission levels at only 1.7 metric tonnes per capita in 2010–2014 (World Bank 2013), on the other, India is being increasingly called upon to assume a leadership role in the international climate change discourse, in keeping with its status as a large country and emerging economic powerhouse. Dubash (2013) posits that India's climate politics have essentially been shaped by the imperatives arising out of its poor and developing nation mindfulness, with a commensurately large burden of development, while it is increasingly being 'forced to grapple' with its status as a large and important emerging power. Dubash places the Indian debate within this paradox of being a poor country on the one hand and a fast emerging global power on the other, simultaneously. However, there are other important nuances that cannot be ignored and are insufficiently captured by this binary.

Hurrell and Sengupta (2012), on the other hand, present a unique western perspective on the debate, while situating arguments in the emerging economy discourse. They argue that emerging economies, given their new found economic dynamism and muscle, have embarked on an irrational path that obfuscates current realities to the visceral and dangerous logic of historic responsibility. They argue that emerging powers 'have failed to recognise or live up to the responsibilities that go with their newly acquired roles', further arguing that their 'governments have all too often proved to be obstructionist and negative' (Hurrell and Sengupta 2012).

We aim to go beyond the narrow confines of the emerging India identity rooted in the geopolitical power discourse that most of these studies situate it in by unravelling the more complex multiple self-identities and attendant factors at play shaping India's response to the global climate change politics.

As the debate stands today, at the macro level, there is a clear gulf in how developed countries and developing countries view their roles and responsibilities. At the fundamental level, there is a 'structural tension' around the division of responsibility and accountability of various countries in terms of the climate change problem (Dubash 2009). The 'responsibility' debate is at the heart of the negotiations, since developing and developed countries see the response to the climate change problem as intrinsically tied to the levels of responsibility of the parties to the UNFCCC (Ghosh 2012), the international treaty that forms the basis for multilateral climate change negotiations today. While developed countries tend to approach the responsibility question from a largely 'techno-managerial' perspective, developing countries, including India, tend to approach it from an 'equity'-centric perspective, framing the problem as one of 'equitable sharing of development space' (Dubash 2009).

The Copenhagen Accord was hard fought and a result of consultations between the US and the then BASIC group of countries (UNFCCC 2009a). India's role in this exercise was critical, and the core Indian concerns were successfully protected, largely attributable to collective bargaining as a bloc and with the co-option of the United States of America.

The first concern was regarding any possible mitigation commitments for developing nations. The Copenhagen Accord did not include any mention of such commitments as they were not envisaged under the Kyoto Protocol (linked to the UNFCCC) for the Non-Annex 1 countries (United Nations Framework Convention on Climate Change 1997). The second was regarding measurement reporting and verification frameworks, where India did not want to be subject to international scrutiny, overriding primacy of its own domestic development goals and frameworks, which it believed to be justified and consistent with the same agreement. The third was regarding means of finance, through which India demanded that developed countries should adhere to the principle of paying for 'agreed full incremental costs' of measures taken by developing countries to mitigate climate change as enshrined in the UNFCCC (United Nations Framework Convention on Climate Change 1993).

At COP-18 convened at Doha in 2012, most of India's core negotiating interests stayed protected. Upadhyay and Mishra (2013) observe that several of India's key concerns were incorporated implicitly at the Climate Change Conference discussions and aligned to the notion of 'equity', which has remained central to the Indian bargaining position despite the conference commitments not measuring up sufficiently on either the issue of financing or transfer of mitigating technologies, two key demands of developing countries championed by the Indian position.

Looking ahead, India faces some tough negotiating choices. At COP-17 held at Durban in 2011, a Platform for Enhanced Action was adopted, which is leading the negotiations towards a new legally reinforced universal protocol in the reduction of greenhouse gases to be made applicable beyond 2020 (UNFCCC 2011). India will have to look inward to assess its development priorities to be able to articulate its red lines in the negotiations and at the same time fulfil its leadership role by proposing effective modalities and architecture for the new instrument. At COP-19 at Warsaw, a decision was taken to urge countries to offer their 'intended nationally determined contributions' by 2015 (UNFCCC 2013). In parallel to the UNFCCC process of negotiations, UN Secretary General Ban Ki-moon invited a range of stakeholders along with the heads of states in an effort to mobilise political will for committing to the climate agreement due by 2015. This 'Climate Summit 2014' noted the absence of Indian and Chinese heads of the states.

The Indian representative, Prakash Javadekar, Minister of Environment, Forests and Climate Change, at the meeting explicitly said that 'by asking for all countries to pledge on equal footing at the New York meeting, the summit is taking a shot at the firewall between developed countries and poor countries. This is not acceptable to us and many others' (Sethi 2014). Amidst the scepticism towards the intention of the Climate Summit 2014, India highlighted the initiatives taken up by the government to improve energy efficiency, renewable energy share and voluntary emission reduction targets and refrained from making any official statements.

Consequently, the pressure was seen building up for India and China to agree to the legal commitment at the COP-21 in Paris. Again, the world looked closely at the development pathway India articulates in consonance with its negotiating strategy going into 2015.

CHAPTER 2

India and Climate Change

Abstract This chapter explores the context in which India finds itself including the development challenge that it faces and the need for energy development. It will provide an overview of the various indicators and sustainability challenges that India faces as a background to developing the identities. It builds the case for and outlines six identities. These identities are used in Part II of the book as a basis for analysis and testing.

Keywords Energy · Development · Growth · Politics

Climate change is certain to bring about profound and unalterable changes to India – environmentally, socially, economically and politically – much like the rest of the world. Surface temperatures are expected to rise on an average by around 1.5 to 2 degrees centigrade in India, with seasons experiencing greater temperature extremes (Ministry of Environment, Forests and Climate Change (MoEFCC) 2012). A majority of India's workforce is still employed in the largely climate-dependent agricultural sector. About 12% of India's cultivated land is flood prone, and 18% is drought prone (India Disaster Knowledge Network 2009). India's persistent food and water security challenges and the projected rise in surface temperatures are sufficient cause for alarm and concomitant domestic executive action. Paradoxically, despite the grim outlook, it is unlikely that climate vulnerability alone will shape how India constructs key policies that will impact on its domestic carbon emissions profile or its negotiating stance at climate forums. It is this very vulnerability that informs its demands for technological and financial intervention, rapid growth, greater social and economic resilience and a larger financial pie to better manage its climate change response.

Since the very beginning of India's involvement in the global climate change debate, even before the Kyoto dialogue in 1992, the interplay between the articulation of its mitigation and adaptation concerns and the more fundamental development concerns were consistently visible (Agarwal and Narain 1991). Due to the inherent linkages between the narratives of climate change and energy use, and in turn the correlation between access to energy and economic development (United Nations Development Programme (UNDP) 2005), the discourse on climate change in India is highly political. The administrative apparatus needs to provide energy at affordable costs to a diverse range of groups – rural–urban consumers, industrial/commercial consumers, excluded social groups, and the politically aware and influential middle class. Given the overemphasis on affordability, combined with the lack of an effective energy infrastructure and the heady urge of irrational populism, taking into account the political ramifications of energy decisions, energy policy in India has largely situated itself around weak economic and environmental logics.

India, along with a group of developing nations, has remained steadfast in using a negotiating position with equity as the cornerstone, maintaining that developed countries have a historical responsibility, based on their past emissions, to facilitate mitigation and adaptation in developing countries, in order for the world to collectively manage the imminent risks posed by climate change (UNDP 2005). Article 4 (paragraph 7, 14) of the UNFCCC incorporates a number of elements mentioned here:

> The extent to which developing country parties will effectively implement their commitments under the Convention will depend upon the effective implementation by developed country parties of their commitments under the Convention related to financial resources and transfer of technology and will take fully into account that economic and social development and poverty eradication are the first and overriding priorities of developing country parties. (UNFCCC 1997)

It is no secret that Indian negotiators bargained hard for the inclusion of this paragraph into the treaty and in many ways, it reflects the collective political position of a large number of developing countries that also pushed for its inclusion (Dasgupta 2012).

India's engagement with the global climate change discourse can be traced to the United Nations Conference on Human Environment (UNCHE) in Stockholm in 1972, where the then Prime Minister Indira Gandhi clarified India's poverty-centric perspective to the world, stating that 'environment cannot be improved in conditions of poverty. Nor can poverty be eradicated without the use of science and technology... The environmental problems in developing countries are not a side effect of excessive industrialisation but reflect the inadequacy of development' (Gandhi 2008).

Mrs Gandhi's approach towards the global climate change problem has informed and influenced the trajectory of India's thought and position on climate change, forming the veritable foundation of India's consistent position over the decades. Twenty years after the Stockholm Summit, at the United Nations Conference on Environment and Development in Rio de Janeiro in 1992, India continued to place emphasis on poverty eradication, environmental justice and equity, financial flows, technology transfer and other mitigation assistance (Ghosh 2012). A further two decades after the 1992 Summit, at the Rio+20 Conference on Sustainable Development in 2012, India again was strongly asserting that poverty eradication remains the greatest global challenge and is the country's foremost priority.

This vocalisation by Indian leadership is today viewed with scepticism and disdain. Some argue that India seeks to hide behind its poverty and deploys this line of argument to absolve itself of responsibility.

Growth and Development Trajectory

The soul of India lives in its villages
M. K. Gandhi

The story of India today is contrasting and at times contradictory. Since 1991, when the pressure of a severe balance of payments crisis led policymakers to open up the economy, create structural efficiencies and liberalise, the country has seen rapid gross domestic product (GDP) growth (Banga 2005). This growth has largely been led by an expansion in the service sector, catering to both domestic and global demands, and a significant increase in per capita consumption. The growth of the service sector has been accompanied by a commensurate decline in the manufacturing and agricultural sectors' share in the GDP pie (Banga 2005). In 2012, India overtook the Japanese economy in terms of purchasing power of parity to become the third largest economy globally. This fast-growing economic prowess has helped India gain significant political weight and influence in the global arena, and its views and actions are increasingly seen to impact and influence the responses of others. However, the risks India's economy faces are systemic and will require both time and effort to mend.

As per the latest census conducted in 2011, India's population has grown by over 235% since 1951 and is projected to continue growing at a steady pace (see Table 2.1).

According to the National Sample Survey (2008–2009), India is still largely rural, with nearly 73% of the population living in villages (see Table 2.2). The latest census conducted in 2011 indicates a substantial demographic shift to urban centres. For the first time in independent India's history, the decadal population growth in urban centres has exceeded that in the rural areas, with around 91 million being added to the urban population as against 90 million to the rural population. Yet, it is important to re-emphasise that the overall distribution of population is still heavily skewed towards rural areas.

In 2012, the nodal agency (at that time and subsequently replaced with the NITI Aayog) for formulating strategies for planned expenditure, the Planning Commission of the Indian government, set the country's poverty line at INR 28.65 per day (approximately USD 0.52) for urban areas and INR 22.42 per day (approximately USD 0.4) for rural areas in terms of per capita expenditure (Planning Commission 2012). Table 2.3 shows the percentages of the total population using the approximated INR 28 as a

Table 2.1 India – Population over the years and projections

Year	Population (millions)
1951	361.08
2011	1,210.21
2026	1,399.83

Source: Census of India 2011, Data points from Government of India (projections from Census 2001)

Table 2.2 India: Rural–urban distribution of population with household size and monthly per capita expenditure (MPCE)

Sector	Frequency		Percentage		Average household size		Average MPCE (Indian rupees)	
	2004–2005	2009–2010	2004–2005	2009–2010	2004–2005	2009–2010	2004–2005	2009–2010
Rural	733,106,178	761,152,796	74.68	72.95	4.88	4.68	558.80	927.7
Urban	248,509,325	282,203,054	25.32	27.05	4.36	4.14	1,052.35	1,785.81
Total	981,615,503	1,043,355,850	100	100	4.74	4.52	683.75	1,159.8

Source: India Datalabs @ ORF, National Sample Survey Household Size Average MPCE, 2004–05 and 2009–10

Table 2.3 Expenditure classes in India (2009–10)

All India	Rural	Urban	Total
Expenditure	% of population	% of population	% of population
<Rs. 28 per day	58.54	22.98	48.92
Between Rs. 28 and 100 per day	40.25	65.54	47.09
More than Rs. 100 per day	1.21	11.49	3.99
Total	100	100	100

Source: National Sample Survey, 2009–10 @ ORF India Datalabs

benchmark. The rural–urban divide is quite evident, with rural dwellers facing stronger financial headwinds.

Though there was a 33% decline in the poverty ratio over a period of seven years (2005–2012), income disparity persists with an 88% (Rajasekhar and Sahu 2006) difference between average monthly per capita expenditure of rural and urban households. These numbers correspond to those reported in earlier studies conducted, for example, by Deaton and Drèze (2002) and Sundaram and Tendulkar (2003). Shukla and Dixit (2007) found an 11-fold difference between the top 10% and the bottom 10% of the income class. The Gini coefficient, a measure of the income distribution of a country used commonly to measure inequality, increased from 0.31 in 1994–1995 to 0.38 in 2004–2005, based on income surveys, which meant that inequality in the country increased despite the rapid rates of GDP growth in this time period (Shukla and Dixit 2007).

This income disparity also manifests itself in inequity in access to resources and basic services (see Fig. 2.1). While on one hand, the declining per capita availability of water in India points towards looming scarcity, on the other, the improving access to water facilities will help the nation meet its human development objectives better. Death rates have decreased and health care costs have gone up. School enrolment ratios show improvement with an increase in the number of schools, yet 25.96% of the population remains illiterate and a significant proportion of the population lacks access to primary education.

Post liberalisation, there has been a structural shift towards the non-farm sector and improvement in labour productivity and incomes. Surprisingly, the compound annual growth rate (CAGR) of employment was negative for all the sectors between 2004–2005 and 2009–2010,

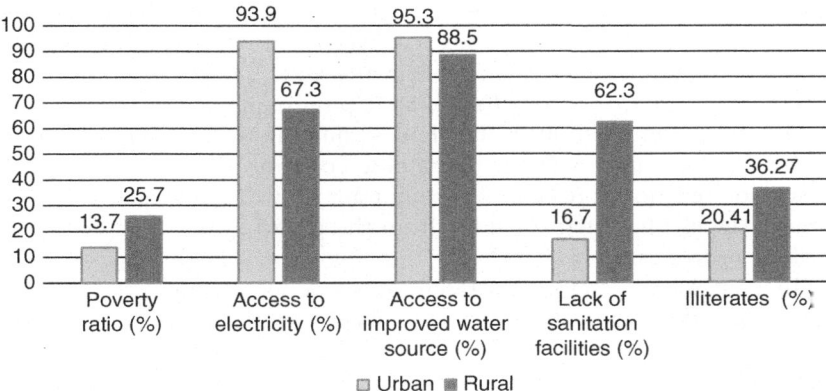

Fig. 2.1 Key indicators of India's development

Source: National Sample Survey (2011) and Planning Commission (2012).
*Data for electricity, water source and sanitation is for 2009–2010; data for poverty ratio is for 2010–2011; data for literacy levels is for 2011–2012

except for the non-manufacturing sector that experienced an enormous growth of 11.02%. Agriculture, with its large share of workforce engagement, showed a sharp decline in elasticity of output growth and negligible productivity growth (Pal and Ghosh 2007). Lack of income diversification, limited access to salaried work and sectoral dependence (especially on low productivity agriculture) are adding to the vulnerability of low-income groups (Planning Commission 2002). Moreover, susceptibility in terms of climate-induced sensitivities is expected to be the highest among low-income groups and rural areas.

A 2007 Greenpeace Report on climate injustice within India questioned the country's persistent demand for global equity and justice in multilateral forums (Ananthapadmanabhan et al. 2007). The findings of the report illustrate that 1% of the Indian population contributes significantly to India's carbon footprint, camouflaged as it was by 99% of the country, bringing the overall per capita emissions below 2 tonnes of carbon dioxide equivalent per year. The report questioned whether India hides behind its poor.

This assertion by Greenpeace, and by other Indian scholars who have also identified substantial variation in the carbon footprint of individuals in

India based on location (urban or rural) (Bhoyar et al. 2014) and/or based on socio-economic classes (Chakravarty and Ramana 2012), implicates the consequential demand for greater equity at home from India. Irrespective of the above, India's historic positioning of the poverty and equity debate internationally did find significant traction among some nations and certainly with the developing countries who shared colonial pasts, poor economic realities and significant growth challenges. Many of them may also have felt excluded from the global governance decision-making processes.

In many ways, therefore, the discussions around environmental responsibility and development choices were shaped by these common experiences. In the days to come these would also become the very basis for these countries in their description and definition of sustainable development. Poverty, unemployment and economic growth would always take precedence over environmental action. This was captured in the statement of Indian Prime Minister Indira Gandhi at Stockholm in 1972, when she stated that poverty remains an obstacle to environmental action, and environmental challenges were a consequence of poor development (Gandhi 2008). Many Indian leaders thereafter have engaged the sustainable development and climate change conversations with a strong agenda that placed poverty and development needs over environmental action.

DEVELOPMENT AND INTERNATIONAL POLITICS

The Group of 77 (G-77) countries is the largest inter-governmental organisation of developing countries at the United Nations (UN), with India as one of its founding members. After the Second World War, many of the countries in the Global South unshackled their colonial associations. This process of breaking away, however, was far from seamless in terms of economic transition. Most of the countries were already locked into fundamentally unsustainable socio-economic growth and development trajectories owing to weak structural foundations and the inheritance of extractive and exploitative policies (G-77 at the UN 2014). At the same time, Bretton Woods Institutions, such as the World Bank and International Monetary Fund, were created to 'advance a liberal international monetary and trading system', with no regulatory body to represent the interests of the Global South (Vihma et al. 2011). This imbalanced narrative led to the vociferous developing country demand for the creation of the United Nations Conference on Trade and Development (UNCTAD), an international organisation, within which the G-77 was formally founded.

While the G-77's origins were tied to the UNCTAD, its mandate fast expanded beyond trade and investment. In the 1970s, environmental concerns and other issues under the purview of UN came under the G-77 mandate. Indeed, the membership of the G-77 also expanded to accommodate more than 130 countries from the Global South. At the 1972 UNCHE in Stockholm, developing countries emphasised the socio-economic development agenda. Vihma et al. (2011) observe that the erstwhile Indian Prime Minister Indira Gandhi expressed and emphasised a 'position that gave precedence to socio-economic development over environmental stewardship, and which viewed them as inherently competing priorities.'

They argue that despite identity being a concept difficult to understand, both analytically and empirically, political posturing around contrasts and tensions in ideological moorings and identity between the Global North and South inform the climate change discourse. Of these, the shared colonial experiences situated in the narrative of injustice and inequity significantly rally the imagination, shaping the responses of the developing world (Vihma et al. 2011).

Since the very beginning of climate change negotiations and the formative days of the coordinating supranational institutions, India has been at the forefront of the climate change discourse and torchbearer for G-77 interests. The fundamental premise of India's positions in climate negotiations has been the differentiation between the roles and responsibilities of developed and developing countries with regard to climate change mitigation. This negotiating position, now an official foreign policy stance, has been steadfast. The then Indian Foreign Minister SM Krishna's observations on climate change and sustainable development at the Preparatory Ministerial Meeting of the Non-Aligned Movement (NAM) at Tehran on 28 August 2012 are instructive in this regard:

> The imperatives of ensuring Sustainable Development and addressing Climate Change are greater than ever before. Recently, we had a successful conclusion of the Rio+20 Conference. It is important that we make all possible efforts to realise the outcome in letter and on the basis of the accepted principle of common but differentiated responsibility and equity. More importantly, we need to firmly anchor the role of the developing world in the decision-making processes (Krishna 2012).

India has persistently made a strong case for the developing world to play a greater role in the decision-making process in the international system since the start of climate change negotiations. As per Sengupta (2012), India's most prominent successes at the international climate change negotiations have been its contributions in terms of equitable ideas and norms to shape the discourse, its success as a coalition builder and defender of its own national interests as well as those of the larger developing world, and its expanded global influence (Sengupta 2012).

Climate change, along with trade, has perhaps mobilised the greatest share of institutional effort and resources from developing countries in the multilateral sphere. Vihma et al. (2011) attribute this to three factors. The first factor is that science has shown that climate change will affect developing countries asymmetrically, despite having contributed relatively little to the problem, creating an intrinsic narrative of 'injustice'. The second is that the imperative of provision of modern energy to all is a key development priority for all developing countries. And third is the financial flows from Global North to facilitate low carbon transition in the developing world with the precondition that developed countries themselves are able to commit to deep and binding cuts in their own emissions in the near future.

Within the G-77, there is a clear, common concern relating to financial and technology transfers and North–South aid flows. Incidentally, the global financial meltdown, which altered global financescapes, has led to some fundamental changes in 'Official Development Assistance' (ODA) policies, defined as the concessional part of official resource flows from the North to the South. Sharan and Kumar (2013) state that there has been a 4% reduction in ODA, measured in real terms, from 2011, and which accounted for less than 0.3% of the donor nations' collective gross national income (Sharan and Kumar 2013).

India has changed dramatically since the 1970s. It is now a two trillion-dollar economy, which is integrated with the world. India is now labelled an 'emerging market economy' with strong potential for sustained economic growth and development driven by a favourable demographic base (O'Neill 2001). A member of the BRICS grouping (Brazil, Russia, India, China and South Africa), India sees itself as having an important leadership role to play in the multilateral sphere. This can be traced to five distinct motivations, which in turn affect the self-perception of Indian stakeholders.

First, BRICS has been envisioned as a grouping distinct to G-77, G-20 or G-7 in terms of its global role. It has been clear from the outset that the ethos and objectives of an informal grouping such as BRICS has to reflect the unique characteristics of the five member countries, each of which has multiple identities and is going through different stages of economic, political and social transition. Therefore, BRICS, by design and motivation, seeks to respond primarily to the inherent overlapping challenges among member states arising out of this transition (Saran and Sharan 2012a).

Second, stakeholders posit that BRICS is not about supplementing, supplanting or competing with any existing structures, institutional or otherwise. Rather, the BRICS serve to magnify the collective impulse of member countries for changing the way the world is managed. Global political and economic governance today still reflects post-Second World War II realties and is still largely driven by the Atlantic countries. Stakeholders hold that BRICS seeks to redefine and reshape this outdated global governance architecture (Saran and Sharan 2013).

Third, member states have invested significant time and money into the BRICS agenda. The grouping is sufficiently mature and has already set up a USD 100 billion New Development Bank – a new multilateral development bank, led by BRICS countries. BRICS is being given shape through its agencies, which will likely change the global development discourse.

Fourth, there is deep acknowledgement that each of the BRICS member states faces immense domestic challenges. No European experience matches the scale and magnitude at which solutions need to be imagined and delivered within each of these nations. Western experiences, rooted in their dissimilar realities, offer no copybook solutions for challenges around health care, education and skilling, infrastructure, and social and behavioural change imperatives faced by these nations. Therefore, BRICS seek to re-contextualise solution sets and work towards a development approach that is suitably flexible, adaptive and unfettered by the successes and failures in distinctly divergent socio-economic contexts (Saran and Sharan 2012a).

Fifth, member states have been engaged with each other, driven by economic imperatives, leveraging between them their inherent competitive advantages. Intra-BRICS trade touched USD 230 billion in 2011 and member states had set an ambitious target of increasing bilateral trade to USD 500 billion by 2015. While the global economy has continued to stagnate, and slower relative rates of growth reflect the contemporary economic reality, the market logic of intra-BRICS cooperation remains inexorable (Press Trust of India 2012).

In addition to the above, a recent trend within BRICS has been the dramatic increase in political content. Political authority across various levels of governance is increasingly becoming closely engaged, with regular interaction between heads of states, political exchanges and experience sharing taking place. BRICS are therefore increasingly likely to leverage the enhanced mutual political understanding such an engagement will likely result in, both internationally and in enhancing productive collaboration among themselves. In the most recent meeting at Ufa, it was noticeable that even Russia, that had remained away from climate consultations until now, seemed eager to participate with the others on environmental policy coordination.

At the Conference of Parties (COP)-15 held at Copenhagen in 2009, India coordinated its responses with a select group of similarly placed developing states, viz. China, Brazil and South Africa, to resist pressure from the US-led Global North (Sengupta 2012). This was to become the basis of the evolution of a new group of developing countries formation called the BASIC (Brazil, South Africa, India and China), which has all the BRICS members with the exception of Russia and that, as was mentioned, may be changing soon. India has relied upon the grouping to take coordinated positions to its benefit in the climate change negotiations.

Since the COP-15, BASIC countries have been meeting on the sidelines of climate change negotiations regularly. BASIC has issued important political statements such as the one which called for the second commitment period of the Kyoto Protocol. As per some observers, BASIC has also been used as a bridge by the member countries to further the cause and voice of G-77 (Hallding et al. 2011).

BASIC is now an evolved group. The 10th BASIC Ministerial Meeting was held in New Delhi in 2012. The countries issued a joint statement, which emphasised the historical responsibility of developed countries and reaffirmed the central principles of equity and common but differentiated responsibility and respective capabilities under the UNFCCC.

Qi (2011) points out that their combined economic muscle, accounting for 15% of the world's GDP, and combined emissions, accounting for 25% of global greenhouse gas emissions, dwarf the rest of G-77 put together, placing the BASIC group in a league of its own. This economic muscle and environmental impact, he argues, affords the BASIC veto-like powers in any negotiation and potential agreement, making it imperative for the rest of the world to engage with them vigorously for any meaningful outcome in the climate negotiations (Qi 2011).

Empowerment at the multilateral level of course does not imply that India will by extension find it any easier to resolve domestic dilemmas. Saran and Sharan (2012b) argue that 'India's macro position on equity at international fora' such as at Rio+20 or the COP Summits 'must be reflected in its domestic resolve to offer energy equitably to its diverse population' (Saran and Sharan 2012b).

Both BRICS and BASIC allow India the ability to help meet its global ambitions, as well as some of its severe domestic challenges, through closer cooperation with countries sharing similar sets of development interests.

However, even as India has pegged the evolution of its international posture and rallied support based on the twin narratives of poverty eradication and development and a degree of South–South solidarity, its central domestic realities that compel it to seek such global leg room are still persistent, the chief among them being energy access and energy insecurity.

Energy, Development and Growth

Energy sufficiency is critical for India to meet its targets on poverty reduction, employment, literacy and human development. The fact that more than 300 million people lack absolute access to modern energy and the country records a power deficit of 12.5% on average highlights why achieving energy security is an immediate goal for India. In fact, India encounters the challenge of securing modern energy supplies for fuelling the economic engine; strengthening infrastructural support systems for transmission and distribution; and building resilience for climate change. The energy challenge in India can be grouped under three main categories: energy access and affordability, energy security and energy production.

Energy Access and Affordability

Access to modern energy, such as electricity, is important for a nation to meet its development goals. UNDP views access to modern energy as the pivotal factor in determining development trajectories of countries. UNDP's 2007–2008 Human Development Report states:

> Access to modern energy services is fundamental to fulfilling basic social needs, driving economic growth and fuelling human development. This is

because energy services have an effect on productivity, health, education, safe water and communication services. Modern services such as electricity, natural gas, modern cooking fuel and mechanical power are necessary for improved health and education, better access to information and agricultural productivity. (Gaye 2007, p. 1)

UNDP describes energy poverty as the 'inability to cook with modern cooking fuels, and the lack of a bare minimum electric lighting to read or for other household and productive activities at sunset' (Gaye 2007, p. 4). In India, the per capita energy consumption is around 481 kWh, less than a fifth of the world's average (Rao et al. 2009). Saran and Sharan (2012b) suggested that the percentage of households using kerosene for lighting in rural areas averages between 30% and 40% for disadvantaged groups, noting that this is a 'striking figure' given that typical kerosene lamps deliver between 1 and 6 lumens per square meter (lux) of useful light, in comparison with Western standards of 300 lux for basic tasks such as reading (Table 2.4) (Saran and Sharan 2012b).

There is little disagreement among scholars that energy poverty is a constraint to economic growth and development. Improving access to modern energy is fundamental to India, achieving its economic and social development goals. It is approximated that a three- to fourfold increase in electricity use in India is required to achieve a Human Development Index ranking of 0.8 (Sant and Gambhir 2012). Noted climate change commentator Prodipto Ghosh (2012) observes that across nations, human development has always been preceded by increase in per capita energy use.

Table 2.4 India and the world – Access to electricity and population

	Total population (in millions)	*Electrification rates (%)*	*Total population without electricity (in millions)*
World	6,692	78.2	1,456
OECD and transition economies	1,507	99.8	3
Developing countries	5,185	72	1,453
India	1,210	66.3	450

Source: International Energy Agency (2011), IEA (2009), Census of India (2011) and Ministry of Environment, Forests and Climate Change (2012)

According to International Energy Agency (IEA), 67% of the rural population has access to electricity as against 94% of urban dwellers (IEA 2009). Arising from this structural bias towards urban areas, India's National Electricity Policy (NEP) (drafted by the Ministry of Power in 2005) deliberately puts higher emphasis on rural electrification (Ministry of Power 2013). The first thrust area outlined by the notification document issued by the Ministry of Power is rural electrification and the rationale given is as follows:

> About 56% of rural households have not yet been electrified even though many of these households are willing to pay for electricity. Determined efforts should be made to ensure that the task of rural electrification for securing electricity access to all households and also ensuring that electricity reaches poor and marginal sections of the society at reasonable rates is completed within the next five years. (Ministry of Power 2013)

The policy outlines the role of the Rural Electrification Corporation of India as the nodal agency to implement the electrification goals of the National Common Minimum Programme of the Government of India – a vision document released by the then coalition government (the United Progressive Alliance) in May 2004, which formed the basis for the review of the then prevailing Electricity Act 2003, and recommended completion of household electrification nationwide in five years (Prime Minister's Office 2013). However, 100% electrification in rural areas is still a distant reality as more recent numbers suggest. According to the status report of Rural Electrification Corporation, about 88.3% of un/de-electrified villages have been electrified across India as of 31 May 2012 (Rajiv Gandhi Grameen Vidyutkaran Yojna 2012).

While the NEP had set the ambitious goal of complete electrification, the inability to achieve targets in time is symptomatic of the implementation gap, and perhaps telling of the nature of the prevalent political economy. However, the suboptimal outcome within the envisioned timeframe was predicted by several analysts. Energy commentator R. Rejikumar states that the policy reads more like a wish list of everything desirable in the Indian power sector, with little deliberation on effective planning and implementation. Considering that electricity is a concurrent subject with federal and provincial governments having equal jurisdiction, he argues that the absence of a clear roadmap would prove debilitating (Rejikumar 2005).

The NEP acknowledges the need to strike a balance between good economics and affordability combined with federal and provincial priorities and states:

> Targeted expansion in access to electricity for rural households in the desired timeframe can be achieved if the distribution licensees recover at least the cost of electricity and related O&M expenses from consumers, except for lifeline support to households below the poverty line who would need to be adequately subsidised. Subsidies should be properly targeted at the intended beneficiaries in the most efficient manner. (Ministry of Power 2005)

Emphasis on subsidies for poor consumers and improved centre–state coordination points to the inherent challenge in scaling up electricity access in India. The commercial non-viability of providing electricity at cheap rates puts the onus of rural electrification almost entirely on the government-run power companies since private parties are unlikely to invest in economically unviable ventures.

The emphasis on affordability and collaborative effort by the state and central governments and the private sector for scaling up output to meet the national energy deficit is the central theme of the Tariff Policy.

While average tariffs have been increasing, it is the commercial and industrial consumers who have borne the greatest and disproportionate burden. The nature of India's political economy is inherently populist. The significant across-the-board subsidy extended to the agriculture sector, on which about 65% of Indian population depends (FICCI-KPMG 2013) bears testimony to this. Grover (2003) argues that it is the industrial consumer who has over the years borne the greatest brunt of thriftless state governments, which have followed populist and unsustainable energy policies and practices. He points to the dramatic reversal in the proportion of subsidising to subsidised consumers between 1970 and 2000 to drive home the point. While in 1971, 74% constituted the subsidising segment and 26% the subsidised segment, in 1999 the respective figures stood at a dramatically reversed 40% and 60%, with consumers paying high commercial tariffs the most burdened (Grover 2003). Dubash and Rajan (2001) observe that electricity subsidies have been used as a political tool since the late 1970s and became 'routine political instruments' thereafter. The political agenda is often veiled (albeit thinly) behind the development agenda in India. Jain (2006) notes that in the case of Punjab, a state where electricity is fully subsidised for the agricultural sector, the subsidy regime in fact reinforces the identity divide between the 'haves and the have-nots' by effectively polarising development gains towards those with large tracts of land and sufficient capital and infrastructure. Jain observes that marginalised farmers are willing to pay 'reasonable user charges' for 'adequate, reliable and quality electricity' that is otherwise unavailable due to the structural distortions created by the subsidy regime, reinforcing the fact that the subsidy regime only serves to accentuate identity disparities – especially income identities. Dubash and Rajan (2001) second this in observing that while 'agricultural power subsidies are typically made on behalf of poor farmers, several studies confirm that the constituencies at stake are so-called "kulak" or landed classes' (Dubash and Rajan 2001).

Energy Production

At present, India consumes 621 million tonnes of oil equivalent (MTOE) ranking India as the world's fourth largest consumer of primary energy. Growth rate of primary energy availability from conventional sources has not kept pace with the growing demand. Supply side challenges can be understood through three basic elements: availability of the resources (domestic reserves and imports), generation capacities and support infrastructure for transmission and distribution.

It is estimated that total domestic energy production of 669.6 MTOE will be achieved by 2016–2017 and 844 MTOE by 2021–2022 (Ministry of Statistics and Programme Implementation (MSPI) 2013). This will only partially meet the requirements – around 71% and 69% of expected energy consumption, in 2016–2017 and 2021–2022, respectively. The balance will have to be met through imports, projected to be about 267.8 MTOE by 2016–2017 and 375.6 MTOE by 2021–2022 (MSPI 2013).

India's energy basket comprises coal with 68% share, followed by natural gas at 12.6%, oil at 11.5%, hydro at 3.3%, nuclear at 2.48% and 1.55% by renewables (MSPI 2013). Coal is clearly the dominant fuel of the Indian economy. The power sector bags the lion's share in raw consumption of coal at 71.3% (403.91 MT), followed by steel and cement industries. Of the total installed electricity generation capacity in India, 62% comprises coal (Ministry of Power 2016).

The Planning Commission of India posits that coal being an abundant resource remains the least cost option and will likely be the primary input source in the foreseeable future, with coal-based generation capacity to be increased to at least 230 GW by 2020 to meet the country's energy security needs (Planning Commission 2011a). However, existing coal plants in the country are based on sub-critical technology, generating relatively high levels of emissions per unit of electricity output, calling for extended focus beyond the resource and capacity generation space into creating efficient power generation infrastructures by additional investments into efficient processes and technologies.

The usage of sub-critical technology in India's coal-fired power stations is causing considerable efficiency losses and environmental damage. Moreover, due to the structural issues in distribution, the aggregate technical and commercial losses in the power sector are immense and show significant variations across states (Fig. 2.2).

The Government of India commissioned a federally administered scheme titled the Restructured Accelerated Power Development and

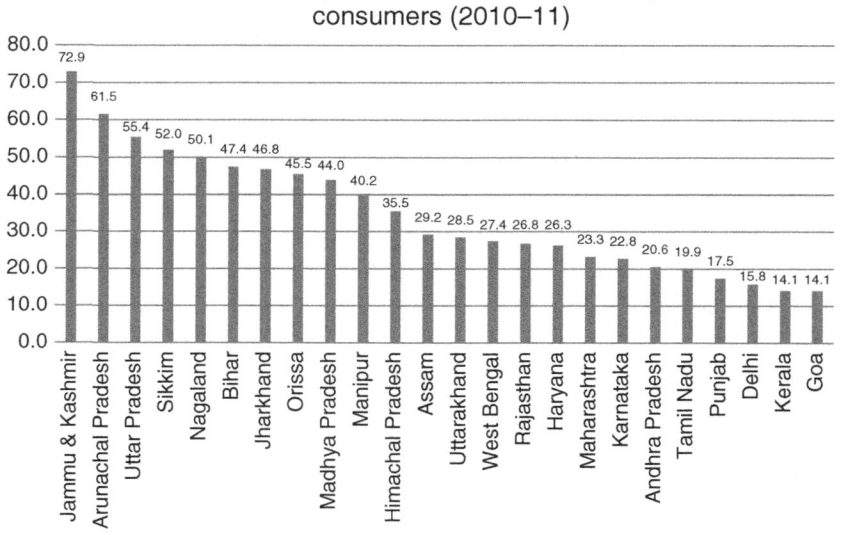

Fig. 2.2 Aggregate technical and commercial losses in the power utilities sector (an illustrative sample of states)

Source: Ministry of Power (2013a)

Reforms Programme in the 11th Five-Year Plan to continue through the 12th Plan to improve sectoral efficiencies and reduce losses in a sustained manner (Ministry of Power 2013a), with the Power Finance Corporation Ltd. (PFC) (a public sector company that invests in the power sector) as the nodal implementing agency.

In his communication to shareholders, PFC Chairman Satnam Singh acknowledges the many challenges the power sector is grappling with. He states:

> In spite of impressive growth in capacity addition and transmission infrastructure, the distribution sector continues to be the weak spot. The financial health of distribution utilities in the country is a matter of concern. The financial losses have shown increase in the last few years. While the utilities should make serious efforts to reduce the Aggregate Technical and Commercial (AT&C) losses there should be adequate tariff to recover the cost of supply. (PFC 2011)

The process of thermal power generation releases high amounts of carbon dioxide and equivalent gases during fossil fuel combustion, making it environmentally detrimental (Sims et al. 2007). However, India's MoEFCC justifies the heavy dependence on coal, upon which thermal power generation is based. It argues that coal being available in large quantities locally at very competitive prices makes better economic sense. It posits that coal-based thermal power is likely to remain the biggest energy source in the foreseeable future (MoEFCC 2012), although the government does acknowledge that overdependence on coal could have adverse ramifications on the nation's energy security (MoEFCC 2012). Government institutions such as India's premier scientific research institute, The Indian Institute of Technology, New Delhi, has expressed concern about the environmental and economic impacts of the overuse of coal. Significantly, in a recent inventory assessment report on thermal power generation, the Institute noted that '0.8–0.9 kg/kWh of carbon dioxide is emitted in Indian power plants' which tend to have a low average net efficiency for power generation (Raghuvanshi et al. 2006).

According to the Coal Ministry, a cumulative total of 285,862.21 million tonnes (MT) of geological reserves of coal has been estimated (proved + indicated + inferred reserves) so far. Non-coking coal, which can be used for thermal generation, accounts for over 87% of total estimated reserves. Coal is found across 11 states in India, with Chhattisgarh, Jharkhand, Odisha, West Bengal, Andhra Pradesh and Madhya Pradesh accounting for the majority of proven reserves. Coincidently, according to the Home Ministry (Internal Security), each one of these resource-rich states is facing civil violence in the form of Naxal insurgency (Ministry of Home Affairs 2010).

However, the development versus environment conundrum looms large in this narrative. Linkages between resource-rich states in India and incidents of Naxal violence are lucidly explained through the paradigm of the multiplicity of socio-economic and political factors by Datt et al. (2009) in a paper on resource federalism in India.

This line of argument discusses the interplay between the multi-functionality of resources and development that lead to three key policy posers. The first is of balancing the interests of the wider constituency (say, the mineral consumer) with those of the more vulnerable local groups that have insufficient agency. The second is seeking to understand and include the interests of different stakeholder groups, particularly the marginalised. And the third is addressing the aspirations to modernity of indigenous (tribal) groups (in areas where minerals are found) while providing them with the option of remaining or being able to return to traditional ways of living, given that people have multiple affiliations that are intrinsic to their lives and that these create contradictions to what they want at different points of times. This, of course, is part of the larger development debate (Datt et al. 2009).

Interestingly, Jairam Ramesh, Environment Minister in the previous government (2009–2014), observes that assertions of right to sustenance visible in different parts of India today, misperceived as conflict between environmental concerns and developmental imperatives, are in reality concerted movements towards securing the very basic of rights, pertaining to livelihood security and determination of developmental choices that impact personal destiny (Ramesh 2010a).

In India, the energy sector was responsible for around 1100.06 million tonnes of carbon dioxide equivalent ($MTCO_2E$) emissions in 2007 (Planning Commission 2011a), accounting for approximately 58% of total emissions. Within it, electricity generation accounted for 719.31 $MTCO_2E$ (Planning Commission 2011b). The climate change model suggests that extreme weather events, resource unavailability and lack of access to factor inputs such as energy will impede the growth of industry, since it would multiply business and functional risks (Sathaye et al. 2006). While resource scarcity issues, changes in tax regimes, global environmental laws and emission trading schemes are expected to affect development in India, a global and local climate change response also presents opportunities (Hoffman and Woody 2008) in terms of investment flows and green market development.

Industrial energy efficiency alone is estimated to represent an investment opportunity of INR 82,575 million (Confederation of Indian Industries (CII) 2013), followed by renewable energy, especially solar and wind power. Other opportunities, such as green buildings, are also on offer as India strives to become a middle-income economy (CII 2013).

The Indian government has set highly ambitious targets for renewable energy in the order of approximately 170 GW from solar, wind, biomass and small hydropower. The share of renewables in the total energy mix of the country increased from 7.8% in 2008–2009 to 12.3% in 2013–2014 (CEA 2015). In terms of the generation capacity, renewables account for less than 2% of the total. However, the untapped renewable energy potential in India is more than 216 GW. The major barriers are capital intensity, immature technology (in some cases), irregularity in supply due to lack of storage technologies, low level of local manufacturing capacities and inadequate infrastructure for evacuation.

Much emphasis has been put on development of the Indian solar energy market. There are numerous government initiatives to support the renewable energy development including the National Mission on Solar Energy – Jawaharlal Nehru National Solar Mission. The Ministry of New & Renewable Energy (MNRE), Government of India, has further declared ambitions to pursue 100 GW of installed capacity from solar energy by 2022. However, there is scepticism over realisation of the target given the administrative lapses, infrastructure deficit, issues of bankability on these projects and financial health of the utilities. Although renewables will be an ideal solution for transitioning to low carbon growth, major political, social and economic restructuring will be required to shift away from fossil fuels. For instance, technically, 100 GW renewable capacity additions replace only 30 GW of coal-based energy. The economics and feasibility of such installed capacities require further calculations and assessments. Business models for renewables have to make economic sense for both consumers and producers, for renewable energy to become a mainstay in the power sector.

Recently, the Pew Charitable Trust listed India as one of the top performing clean energy economies in the world. With 54% increase in growth over the previous year, India ranked sixth with investments exceeding $10.2 billion (Trusts, Charitable & Finance 2010). Of the total investments in India in the clean energy sector in 2011, wind energy accounted for USD 4.6 billion, closely followed by solar energy at USD 4.3 billion (Pew Charitable Trust and B.N.E. Finance 2010). In addition, there is tremendous potential for energy efficiency savings through demand side management (see Table 2.5).

However, challenges to renewable energy development in India are numerous. The subsidy regime, which caters to the agricultural sector, also makes the usage of renewable power sources practically unviable for the distribution companies that are mandated to purchase proportions of total power in the form of renewable energy. The Renewable Purchase Obligation (RPO) for the distributors varies between 2.1% and 10% for non-solar generation (Consolidated Energy Consultants Limited 2011).

The correlation between the imperative for capacity addition and the specialised tariff regime for renewable technologies such as solar presents a real cost dilemma (at current technological levels). RPOs are the key policy drivers for promoting solar power (Ministry of New and Renewable Energy 2010).

The Indian government also faces several constraints in financing a growth trajectory based on renewable energy. Consistently rising levels of revenue expenditure account for increasingly large proportions of total expenditure outlay, given the ever-increasing spending on the many temporary stimulus programmes that cater to various socio-economic groups and other political push and pull factors.

It is an economic reality that future growth in the renewables and energy efficiency segments will be based on commercial logic and not on fiscal incentives and subsidies alone. The limited ability of the Indian budget to finance clean energy has not prevented the impressive growth of the renewable sector, and, India is ahead of Japan, China and the United States on procurement of renewable energy on a per capita basis

Table 2.5 The scope for increasing end-use efficiency through demand side management in India

Sector	Potential (%)
Industry	10–25
Lighting	30–35
Commercial buildings	50
Agriculture	40–45

Source: Bureau of Energy Efficiency 2008

(Saran and Sharan 2015). Ironically, to continue to develop this segment, India may have to allow for external engagement that would be conducive to this and may contradict the development and growth space the country has so assiduously protected since 1972.

Energy Security

India is concerned about increases and unforeseen variability in global energy prices as it poses a threat to energy security and the current account. Significantly, in the third quarter of 2012–2013, India's current account deficit surpassed 4.8% of GDP. This led to a relatively sharp depreciation of the rupee, as foreign investors began to pull portfolio investments out of the country. The signalling value of a weak current account became evident to policymakers, and a slew of commentary in the media regarding the precarious fiscal health of the economy added to the domestic concerns.

According to the Directorate General of Commercial Intelligence and Statistics, India, crude oil and refined products account for nearly a third of total imports of principal commodities (Ministry of Commerce and Industry 2014). As a result of this heavy overdependence on crude imports in particular, the Indian economy is vulnerable to external shocks. Since India imports around three-fourths of its crude oil requirement, such fluctuations adversely impact India's balance of payments and, concomitantly, its economic development.

India imported 37% of its oil demand in 1990. However, since then, the share of imported oil to meet domestic demand has gradually risen. Oil imports rose to around 2.7 million barrels per day, catering to 75% of demand in 2012, and is expected to rise to 6.8 million barrels per day catering to 92% of anticipated demand in 2035 (International Energy Agency 2011).

In the case of natural gas, around 30–40% share of current supplies to the Indian gas market is based on short-term contracts or spot market. With increase in gas demand from countries like Japan and South Korea, especially after the fallout of the nuclear reactor disaster in Japan in March 2011, prices in the spot market are expected to remain high. Delays and limited liquefaction capacities available globally will further add to the competition for spot gas. This may be challenging for India to secure gas at a competitive price.

Imports account for 21% of total Indian gas supply and are expected to increase to almost 35% by the end of the 12th Five-Year Plan

(2017), in light of the falling domestic supply. The actual utilisation of liquified natural gas (LNG) re-gasification capacity will also depend on the price developments in the international LNG market (Ministry of Petroleum and Natural Gas 2013).

Owing to its large reserves, India has historically imported small volumes of coal. In 1990, India imported 3% of its coal requirements and all imported coal was coking coal (Ahn and Graczyk 2012), which is in short supply in India. However, by 2012–2013 the absolute volume of India's coal imports reached 146 MT, an increase of around 45% over imports in the preceding year. According to the review of the 11th Five-Year Plan, nearly 15% of the overall coal demand in India in 2011–2012 was met through imports. Of these, non-coking coal accounted for nearly 70%, while the remaining third was coking coal (Lok Sabha 2013).

Domestic production of oil, gas and coal has failed to keep up with growing demand. While overall energy production grew from 378 MTOE in 2001 to 551 MTOE in 2011, at a CAGR of 3.65%, domestic demand far outstripped this rate of growth at 4.5% over the same time period. This has led to weak current account balance and has placed severe strain on the exchequer, as oil imports account for the largest proportionate share of resource imports in the entire imports basket of the country. In 2013, India had a negative balance of trade of USD 131 billion largely on account of growing energy imports.

The technological and financial imperatives for improving energy and operational efficiencies, as well as calibrating policy planning to minimise environmental damage, are clear. India has the difficult task of increasing generation and managing costs to serve its low middle-income economy identity, while unfettered growth will likely remain the cornerstone of economic policy. All of these impulses are visible in the different energy identities projected by India.

Delineating the Identities

According to the United Nations Development Programme 'the human development approach, developed by the economist Mahbub Ul Haq, is anchored around Amartya Sen's work on human capabilities, often framed in terms of whether people are able to "be" and "do" desirable things in life' (HDRO Outreach 2015). Key aspects of this approach include lifeline energy infrastructure such as housing and access to meaningful employment opportunities. Alkire (2002) defines the myriad dimensions of human development as 'non-hierarchical, irreducible, incommensurable and hence basic kinds of human ends' (Alkire 2002). It is from this perspective that the concept of development generally, and the centrality of lifeline energy as an indicator of development specifically, has been employed in this research in order to develop the important Indian identities.

Key observations from the multitude of India's 'lived experiences' as posited by social scientists and philosophers and from the above discussion reveal six distinct identity groups relevant to this study.

The fact that rural India stands disadvantaged in terms of access to energy, and that most of the country's energy access issues pertain to grossly underdeveloped infrastructure and supply side inequity in India, implies that the *rural identity* will likely be a critical component of India's self-identification. While poverty is prevalent in urban centres as well, it is most often vocalised through the grammar of agrarian and rural deprivation. This is deeply enmeshed in the democratic clout that rural India enjoys and is messaged to appease and appeal to this segment of electorate.

India is a resource-scarce country, with an ever-widening demand supply gap and this inequity is very pronounced in the energy domain. There are macro-deficits, which are implicated by the huge dependence on imported fossil fuels and then there is the micro-reality of exclusions where millions of people are still to benefit from electricity or commercial energy sources. Energy supplies and the imperative for securing them is a definite priority and this in turn manifests as India's *energy security identity*.

Similarly, the country has to contend with the rapidly growing energy needs for its industrial growth and job creation to respond to its young population's needs. This *industrial identity* will be an important locus of identification.

However, the demand for cleaner and environment-friendly technologies to replace extant carbon-intensive technologies in critical sectors such as power, manufacturing and chemicals and fertilisers presents an unprecedented economic and growth opportunity. Indian renewables sector has grown impressively in the past few years (Khare et al. 2013) and increasingly a business case exists for the low-carbon business models (Shukla and Chaturvedi 2013). This will inform India's *entrepreneurial identity*, another key pivot of identification.

Energy narratives coincide with large development aspirations and these exist in a negotiated space, between domestic priorities, global aspirations and supranational obligations. Global stakeholders see India as either part of the loud oppositional voice within the bloc of G-77 countries, the *developing country identity*, or as part of a weightier bloc of BRICS and BASIC countries, a group of fast emerging economies bringing commensurate weight to the negotiating table (Dubash 2013), giving rise to *emerging country identity*.

The structural asymmetries in India's development trajectory and its participation in global development politics are manifested through these identities. These, in turn, are informed by a narrative of populism, which influence policy, and indeed the spectrum of choices available within the development landscape to negotiate at the climate change high table.

That India needs to respond to all these contemporary self-imaginations of the nation is increasingly apparent from its responses in the past, faulted for being self-contradictory at times and feeble at other instances by some stakeholders that engage with India as a nation. The multiplicity of its responses can be explained if the nation were to be reimagined through the narrative of multiple identities that have been uncovered in the preceding discussion. Those following the recent development of the Indian state are familiar with some of this and yet they seek 'unity' confused with 'singularity' in its approach. India, while acting as a global or regional power with commensurate responsibility, will also need to respond to domestic inequities and challenges and some fundamental developmental imperatives, such as universal energy access, water, food and social infrastructure. This diverse basket of needs at home cannot be ignored, even while at international forums.

This multiplicity of *identity* alongside multiple expectations, varied perceptions and diverse responses on managing the nexus between development, energy needs and policy to respond to climate change presents a challenging scenario for policymakers. And the process in which responses are shaped and influenced will likely determine the carbon trajectory of not just the Indian economy, but of possibly the entire developing world.

The central hypothesis we present is that India struggles to tackle energy needs, development goals and climate change imperatives together due to the differing drivers it needs to accommodate when formulating responses to these challenges. One of the ways to test this hypothesis would be to uncover the presence of one or more of these identities in the Indian climate change discourse generally, and in the official Indian position at climate change negotiations specifically. In other words, which identity or identities are seen to dominate the Indian discourse on climate change and its positions at international climate negotiations?

This will also allow us to respond to some of the other pertinent queries including

1) Does India hide behind its poverty or does it hide its poverty at global climate forums?
2) Is the lure of the 'global high table' distracting India from its core development objectives?
3) How is the interplay between official positions, media reportage and Indian identities evolving?

These six self-identities *rural, energy security, industrial, entrepreneurial, developing nation* and *emerging nation* are sought to be uncovered in Indian climate discourse with the help of media content analysis of national newspapers (online), semi-structured interviews of elite stakeholders and by analysing speeches made by Indian officials, in the following chapters.

Any attempts to decipher India as a single country, single identity or driven by linear needs and ambitions will result in futile and meaningless understanding of Indian complexity, especially on an issue as complex and all-encompassing as climate change. It is argued that decoding India to its rational development constituents as has been attempted in this study constitutes a pragmatic and effective tool for understanding India on this subject and will allow for a more meaningful and productive engagement with this country in the negotiations that are underway.

PART II

Uncovering Indian Climate Identities

CHAPTER 3

Media Coverage in India

Abstract This chapter explores how the various identities are expressed within popular media in India. It is based on a media content analysis that explored 200 newspaper articles before, during and after key negotiation processes (namely Conference of the Party meetings of the United Nations Framework Convention on Climate Change).

Keywords Content analysis · Newspapers · Media

Media, called the fourth pillar of democracy, plays a crucial role in providing information and shaping public opinion. As American playwright Arthur Miller once stated, 'a good newspaper, I suppose, is a nation talking to itself' (Miller 1961), and this holds true for the Indian media. In India, V. K. Ramachandran, former deputy editor, The Hindu Group, noted that 'owing to the low levels of literacy, the dissemination of information by means of the written word goes much deeper. This has important implications for the quality and depth of public opinion and of participatory democracy in the State' (Ramachandran 1996).

Until the early 1980s, English remained the most popular language for newspapers in India. The fact that only 2% of Indians understood the language as compared to the 40% who could read Hindi did not change this statistic (Malhotra 2008). By the early 1990s, along with the Hindi newspapers, a lot of other regional language newspapers started getting registered, steadily increasing their share of circulation. Today there are more than 28 languages in print. As the trend depicts, Hindi takes a lion's share, followed by English and bilingual language publications (Table 3.1).

When comparing the Hindi and English print media on their reportage on issues of environment, health, politics, science and technology and others, Dr Meenu Kumar conducted a comparative study in 2013. Through content analysis, she found that the English newspapers cover a lot more topics on science and technology as compared to their Hindi counterparts. The study further demonstrated that environment is the second most important topic in both languages, although higher in English (31.78%) than in Hindi (21.78%), only preceded by health issues (Kumar 2013). She concluded by ranking the English newspapers higher than the Hindi ones in terms of space devoted to environmental issues and science and technology, number of international sources used to publish the material and the quality of the publication. This study is important as it confirms the larger participation of the English media on issues that this book seeks to engage with. Additionally, while many studies confirm that regional media has larger circulation, the study conducted by Kumar (2013) would suggest that news manufacturing on issues of climate change and international relations would still reside with the mainstream English newspapers.

The early twenty-first century saw the exponential growth of Internet users in India, recording a growth rate of 137% in 2002 (Internet Live Stats 2015). India now has the third largest Internet user base, after China

Table 3.1 Languages of the Indian newspapers (2013–2014)

Language	Concentration of publications
Hindi	11,184
English	1,889
Bilingual	812
Urdu	1,443
Tamil	231
Marathi	600
Bengali	179
Punjabi	149
Other regional languages (18)	3,012

Source: Press in India Report, Ministry of Information and Broadcasting, Government of India (2014)

and the United States with 302 million Internet users. The number was expected to rise to 354 million by the end of 2015, as per the 2014 Report by Internet and Mobile Association of India and IMRB International. This phenomenon is not limited to urban areas. The growth rate of Internet users in urban and rural India is comparable – 29% in urban areas and 30% in rural areas for the year 2013–2014 (Internet and Mobile Association of India 2014). Internet access through smartphones and other handheld devices have accelerated the transformation of the digital landscape. Additionally, rapid socio-economic progress, a young demographic and the penetration of information and communication technology will further drive this digital revolution in India.

This chapter presents the results of a media content analysis (MCA) on a range of (online) articles. Articles were chosen from five daily newspapers, namely *The Hindu, The Economic Times (ET), Hindu Business Line (HBL), Times of India (TOI)* and *Hindustan Times (HT)*. These newspapers are the most circulated dailies in their respective segments, and their selection meant the analysis included reportage across the entire ideological spectrum as well. *The Hindu* (broadsheet) and *HBL* (business daily) are perceived as leaning to the left on most political and economic debates. The *TOI* (broadsheet) and the *ET* (pink paper) are perceived to be more pro-market and pro-industry, and the *HT* (broadsheet) may loosely be classified as a centrist paper. A note of caution must be tagged with such definitions as this exercise is at best approximate and subjective. It may be argued that the entire Indian debate itself has left leanings and in that sense, these definitions are a relative positioning of these papers as opposed to following global definitions of ideologies.

Two key events were chosen as a focus for the media analysis. First, Conference of Parties (COP)-15 in Copenhagen in 2009 because of its significance, certainly in the run up to it, in the climate change negotiation process. Additionally, COP-18 in Dohain 2012 was chosen as it became a key focus for building a post-Copenhagen consensus and was therefore a bridge from the old political process. The search term 'climate change' was used to discern articles containing the phrase from the online archives of the five dailies. A one-month period before and after COP meetings held in 2009 and in 2012, respectively, were analysed.

Out of the total sample set of 200 articles for the two years, 100 articles were selected for 2009 and 2012 using randomiser software in order to remove biases. These media reports were further categorised under 'before', 'during' and 'after' the COP based on the dates of the COPs.

It is important to mention that the findings in the following section are presented in the form of percentage of the articles that had discussed a particular identity and shows the distribution of the argument.

INDIA'S CLIMATE CHANGE IDENTITY BEFORE COP-15 IN 2009

Out of the 100 articles analysed, 47% of the articles belonged to the period before COP-15 (7 December 2009). The responses of these articles are mentioned in Table 3.2.

Out of 47% of the articles, the *entrepreneurial identity* received maximum attention, with 34% of the articles mentioning green opportunities and a mood to shift towards low-carbon economy. About 66% of the articles did not mention any part of this identity.

The *industrial identity* was captured by the second highest number of articles (29.8%), blaming industries and unplanned growth for excessive carbon emissions. About 70.2% did not touch upon any aspect of the *industrial identity*.

The *rural identity*, linking poverty, dependence on agrarian economy, subsidies and energy affordability, got a mention in 27.7% of the articles analysed before COP-15 commenced. About 2.1% of the articles reported vagueness in their arguments and depiction of this identity and 72.3% of the articles did not refer to this identity at all.

About 27.7% articles touched on India's *emerging nation identity*. India's BASIC (Brazil, South Africa, India and China) membership and a potential role to be the leader in the climate change negotiations along with China

Table 3.2 Media content analysis for reportage on climate change before COP-15 (before 7 December 2009)

Identities	Yes (%)	No (%)	Ambiguous (%)	Not mentioned (%)
Entrepreneurial	34.0	0.0	0.0	66.0
Industrial	29.8	0.0	0.0	70.2
Rural	27.7	0.0	2.1	70.2
Emerging country	27.7	0.0	0.0	72.3
Developing country	21.3	0.0	6.4	72.3
Energy security	10.6	0.0	2.1	87.2

received positive responses with no uncertainty in the arguments. However, the percentage of 'not mentioned' category remained high at 72.3%.

About 21.3% of the articles portrayed India as a developing nation, either grouped with Group of 77 countries (G-77) or the Global South. Responses to this identity were met with scepticism and contradictions with 6.4% of the articles doubting India's *developing nation identity* claim. Articles not mentioning any aspect of this identity were similar to the *emerging nation identity* at 72.3%.

The *energy security identity* scored the lowest with only 10.6% of the articles capturing the debate between climate change and energy security challenges. Also, 2.1% of the articles showed doubt in linking energy-related concerns with climate change. The 'non-mentioned' category for this identity had 87.2% of the articles.

India's Climate Change Identity During COP-15 in 2009

Media reports analysed during the period of COP-15 2009 from 7 to 18 December 2009 were 36% of the total articles. The responses to the questions asked regarding India's climate change identities are mentioned in Table 3.3.

Out of 36% of the total articles that belonged to the period between 7 and 18 December 2009, the *developing nation* and *emerging nation identity* of India bagged lion's share with 38.9% of the articles grouping India both as a member of G-77 and hyphenating it with China. The narrative around the *emerging nation identity* was noticed to be a stronger counter to western position in climate negotiations. While the ambiguities soared high in depicting the *developing nation identity* at 8.3%, none were reported for the *emerging nation identity*.

Table 3.3 Media content analysis for reportage on climate change during COP-15 (7–18 December 2009)

Identities	Yes (%)	No (%)	Ambiguous (%)	Not mentioned (%)
Developing nation	38.9	0.0	8.3	52.8
Emerging nation	38.9	0.0	0.0	61.1
Industrial	30.6	0.0	0.0	69.4
Entrepreneurial	22.2	0.0	2.8	75.0
Rural	19.4	0.0	0.0	80.6
Energy security	8.3	0.0	0.0	91.7

The *industrial identity* received affirmative responses in the media reports at 30.6% with no ambiguity. As compared to the media reportage before COP-15, a slight increase was noticed. About 69.4% of the articles did not mention any aspect of the *industrial identity*.

The *entrepreneurial identity* recorded a drop of almost 12% from the reporting before COP-15. Only 22.2% of the articles mentioned innovation and demand related to green markets, energy efficiency and low-carbon pathways; 2.8% of the articles expressed doubts over linking climate change and *entrepreneurial identity*; and 75% of the articles refrained from making any reference to this identity.

A fall of nearly 8% was observed when compared to the media reports printed before the COP-15 on the *rural identity*. During COP-15, *rural identity* gained much less traction at 19.4%. No ambiguities were reported, and 80.6% of the articles fell under the category of 'not mentioned'.

Contrastingly, the *energy security identity* also saw a decline in the percentage of the articles mentioning energy security obligations and climate change impacts (8.3%) with no vagueness in their reportage. In comparison, the reportage on this identity before COP-15 was higher at around 10%; and 92.7% of the articles did not refer to any characteristic of the *energy security identity*.

INDIA'S CLIMATE CHANGE IDENTITY AFTER COP-15 IN 2009

The media reports analysed from 18 December 2009, after the period of COP-15, were only 17% of the total articles. The responses to the questions posed to these articles are mentioned in Table 3.4.

Out of the total 100 articles, the 17 articles that reported on climate change negotiations after COP-15 showed startling differences from the

Table 3.4 Media content analysis for reportage on climate change after COP-15 (18 December 2009 onwards)

Identities	Yes (%)	No (%)	Ambiguous (%)	Not mentioned (%)
Rural	35.3	0.0	0.0	64.7
Entrepreneurial	35.3	0.0	0.0	64.7
Emerging nation	35.3	0.0	0.0	64.7
Developing nation	29.4	0.0	0.0	70.6
Industrial	11.8	0.0	0.0	88.2
Energy security	0.0	0.0	0.0	100.0

reportage during COP-15. The *rural identity* shot up with a whopping 35.3% of the media reporting in agreement to the identity and no ambiguities were found in these results. This number is much higher than the number of articles reporting the rural discourse during COP-15. About 64.7% of the articles refrained from mentioning any aspect of this identity.

The *entrepreneurial identity* received similar attention to the *rural identity*. About 35.3% of the articles mentioned innovating measures and processes to improve energy efficiency, develop and absorb green technologies and the urgency to focus on low-carbon pathways. No ambiguities were recorded, and 64.7% of the articles did not touch upon this identity.

With similar percentage as *rural* and *entrepreneurial identity*, the *emerging nation identity* made its mark with 35.3% of the articles hyphenating India with China or recognising India's role as a global manager. No ambiguities were found in the results, and 64.7% of the articles did not mention India's *emerging nation identity*.

India's *developing nation identity* noticed a drastic fall from the reportage during COP-15, and 29.4% of the articles mentioned India's developmental needs and focused on common but differentiated responsibilities (CBDR) and equity principles, as compared to 38.9% during COP-15. However, increase in attention given to the *developing nation identity* can be noticed in comparison to reportage before COP-15 – 21.3%. No articles showed ambiguity and 70.6% of the articles did not mention it at all.

India's Climate Change Identity in 2009

The combined result of the MCA for 2009 is presented in Table 3.5. These results represent reportage of 100 articles including reporting done before, during and after COP-15.

A significant inclination towards India's *emerging nation identity* at 33% with no articles showing ambiguity in the sample texts analysed is evident. These articles were fairly direct in their usage of the term 'emerging', others hyphenated India with China and emphasised on the role it plays in the BASIC alignment. Out of the total 100 articles reviewed for 2009 reportage, about 67% articles did not mention *emerging nation identity* at all.

The second most significant identity flagged by MCA 2009 was the *entrepreneurial identity* with 30% of the articles mentioning the capacities

Table 3.5 Media content analysis for 2009 reportage on climate change (COP-15 aggregate)

Identities	Yes (%)	No (%)	Ambiguous (%)	Not Mentioned (%)
Emerging nation	33.0	0.0	0.0	67.0
Entrepreneurial	30.0	0.0	1.0	69.0
Developing nation	29.0	0.0	6.0	65.0
Industrial	27.0	0.0	0.0	73.0
Rural	26.0	0.0	1.0	73.0
Energy security	8.0	0.0	1.0	91.0

and capabilities of the nation states in building green economies and technological innovations, and reducing energy intensity. Only 1% of the articles reflected uncertainty in linking climate change with innovation and entrepreneurship and 69% of the articles did not mention it at all.

The *developing nation identity* bagged the third position and was reported by 29% of the articles. These articles grouped India with the G-77 order, adhering to the principle of CBDR and equity. Considered as a part of the Global South, India was noticeably stated as a developing country. However, another 6% of the articles mentioned this identity with slight vagueness. While these articles obtusely mentioned one or two characteristics of the *developing country identity*, it was difficult to infer the explicit link. Hence, they were labelled as ambiguous.

About 27% of the total texts analysed for 2009 MCA linked *industrial security identity* with climate change. These articles basically held industrialised nations for high-carbon emissions. More specifically, emissions from power, infrastructure and transport sectors were also blamed for emitting greenhouse gases affecting the environment. Claims made by the articles were certain and showed no ambiguity. However, a large number of the articles did not refer to this identity (73%).

The second last scorer was *rural identity* at 26%. The coders looked for cases where poverty, affordability and vulnerability to climate change were stated as the major challenge for climate change mitigation/adaptation efforts and negotiating position. Dependence on agriculture, food security and agricultural subsidies was considered as part of the *rural identity* during analysis of media reports. Out of the 100 articles, 73% of the articles did not mention *rural identity* of any kind. Surprisingly, 1% of the articles put doubt on the science of climate change predicting adverse impacts on agriculture. This indeed requires a special mention. In a written reply to Rajya Sabha, the then Environment Minister, Jairam Ramesh, categorically said that, 'The studies conducted by the Indian Council of Agricultural Research (ICAR) do not reveal confirmed findings about the adverse impact of climate change on Indian agriculture.'

The *energy security identity* showed a startling lack of consideration in the media reportage in 2009. Only 8% of the articles mentioned vulnerability due to heavy reliance on coal and the constraints of the power sector, given the double-digit growth ambitions and growing urbanisation. While only 1% of the articles showed ambiguity in the results, 91% did not link energy security with climate change concerns.

India's Climate Change Identity Before COP-18 in 2012

Out of the 100 sample texts of 2012, articles reporting on climate change before COP-18 in Doha were counted to be 29%. Following are the results from their content analysis (Table 3.6).

Out of the total 29 articles reported before COP-18, with 37.9% of the articles, the *entrepreneurial identity* dominated the reportage. These articles associated climate change with the emergence of green market, clean technologies, alternative forms of energy and innovations related to energy conservation. No ambiguities were linked to the results, and 62.1% of the articles did not mention the *entrepreneurial identity*.

The *emerging nation identity* received second highest attention with 24.1% of the articles referring to India's proactive role at the climate summit and its alignment with the Chinese position. However, 3.4% of

Table 3.6 Media content analysis for 2012 reportage on climate change before COP-18 (before 26 November 2012)

Identities	Yes (%)	No (%)	Ambiguous (%)	Not mentioned (%)
Entrepreneurial	37.9	0.0	0.0	62.1
Emerging nation	24.1	0.0	3.4	72.4
Industrial	20.7	0.0	0.0	79.3
Developing nation	17.2	0.0	6.9	75.9
Rural	6.9	0.0	0.0	93.1
Energy security	6.9	0.0	3.4	89.7

the articles recorded vagueness in their reference to India's emerging discourse.

About 20.7% of the articles asserted the link between the *industrial identity* and climate change. Carbon emissions due to industrial development and growth were directly pointed out. Mostly the indication was towards the industrialised or western nations and their historical responsibility owing to larger contribution towards carbon emissions currently present in the atmosphere. No mention of this identity was found in 79.3% of the total texts studied.

Grouping India with G-77 and the Global South, 17.2% of the articles reported on India's *developing nation identity*. Interestingly, the highest ambiguity was noticed in reportage for this identity, and 6.9% of the articles demonstrated elusiveness towards categorising India as a poor country – seeking funding.

Surprisingly, the *rural identity* showed a stark difference in reportage from 2009 MCA results. Reportage on this identity before COP-15 in 2009 was 27.7%, which dropped dramatically to 6.9% in 2012 before COP-18. No ambiguities were recorded in the sample texts and 93.1% refrained from reporting any aspect of this identity.

The *energy security identity* gathered least amount of traction in media reportage before COP-18. Only 6.9% of the articles linked dependence on fossil fuels or energy scarcity to the climate change debate. About 3.4% of the articles were sceptical in establishing this link and 89.7% of the articles did not mention it at all.

INDIA'S CLIMATE CHANGE IDENTITY DURING COP-18 IN 2012

Out of 100 articles, 43% of the articles of the sample set belonged to the period during COP-18 held in Doha from 26 November to 7 December 2012. The responses from these articles belonging to India's five mainstream English online newspapers are given in Table 3.7.

Out of the 43% media reports, India's *emerging nation identity* dominated the media space (39.5%). The percentage is comparable to 2009 MCA results (38.9%). However, while in 2012 ambiguities in the media reports were found in 2.3% of the articles, there were none in 2009.

The MCA for the year 2012 showed slight increase in media reportage for the *industrial identity* (34.9%) – upturn by 4.3% from 2009 MCA of COP-15. There were no ambiguities in the results.

Table 3.7 Media content analysis for 2012 reportage on climate change during COP-18 (26 November–7 December 2012)

Identities	Yes (%)	No (%)	Ambiguous (%)	Not mentioned (%)
Emerging nation	39.5	0.0	2.3	58.1
Industrial	34.9	0.0	0.0	65.1
Developing nation	34.9	0.0	11.6	53.5
Entrepreneurial	32.6	0.0	2.3	65.1
Rural	25.6	0.0	0.0	74.4
Energy security	11.6	0.0	0.0	88.4

The *developing nation identity* gained responses similar to the *industrial identity*, with 34.9% of the articles discussing India's alliance with G-77 and highlighting developing nation characteristics. Nevertheless, ambiguities noticed for this identity were highest among all other identities at 11.6%. This trend is similar to 2009 MCA (of COP-15); however, vagueness in reportage for the *developing nation identity* in 2009 was lower than in 2012.

The *entrepreneurial identity* experienced a dip in reportage from before COP-18 started, and 32.6% of the articles linked climate change with low-carbon pathways and energy conservation measures and innovations. Unlike the reporting before COP-18, 2.3% of the articles demonstrated uncertainty in their arguments towards this identity.

The MCA showed that 25.6% of the articles discussed the *rural identity* with absolutely no ambiguity. Continuing agrarian dependency, high subsidies regime and vulnerabilities related to agriculture were reported as a major cause of concern, viz. climate change in these articles. Like the previous identity, the increase in attention from 2009 reportage (19.4%) for this identity is noteworthy.

During COP-18, only 11.6% of the articles acknowledged the *energy security identity* and identified soaring energy demands and heavy dependence on coal to be important aspects that need to be considered while planning for adaptation and mitigation measures.

INDIA'S CLIMATE CHANGE IDENTITY AFTER COP-18 IN 2012

Out of 100 articles, 28% of the articles from the sample set analysed were logged from the period after COP-18 was held in Doha, 7 December 2012 onward. The responses from these articles belonging

Table 3.8 Media content analysis for 2012 reportage on climate change after COP-18 (7 December 2012 onward)

Identities	Yes (%)	No (%)	Ambiguous (%)	Not mentioned (%)
Entrepreneurial	39.3	0.0	0.0	60.7
Rural	32.1	0.0	0.0	67.9
Emerging nation	28.6	0.0	3.6	67.9
Developing nation	25.0	0.0	17.9	57.1
Industrial	14.3	0.0	0.0	85.7
Energy security	14.3	0.0	3.6	82.1

to India's five mainstream English online newspapers are given in Table 3.8.

Out of 28% of the total articles after COP-18, the *entrepreneurial identity* dominated the media space with 39.3% articles stating the need to develop green market, invest in clean technology and focus on low-carbon economy. This marks a significant increase in reportage from during COP-18, where the characteristics of this identity were mentioned by 32.6% of the media texts.

The *rural identity* scored the second highest at 32.1%. These articles linked agrarian vulnerability to climate change due to high poverty ratio and lack of livelihood diversification. There were no ambiguities in these results. This is a trend dissimilar to the 2009 media reportage, where 35.3% of the articles stated characteristics of the *rural identity* and impacts of climate change.

India's *emerging nation identity* attracted responses from 28.6% of the articles after COP-18. There is a marked 11% drop from the reportage during COP-18 for this identity.

Similar to the *emerging nation identity*, India's *developing nation identity* noted reduced attention at 25% as compared to 34.9% during COP-18. The ambiguities soared high and were of the order of 17.9% – higher than any other identity.

The *industrial identity* was at par with the *energy security identity* in reportage after COP-18 at 14.3%. While the decline in reporting is fairly evident and extraordinary for the *industrial identity* (drop by 20.6%), the *energy security identity* demonstrated a slight increase from reportage during COP-18 (up by 2.7%). However, while there were no ambiguities in the results during COP-18, 3.6% of the texts analysed after COP-18 showed uncertainty in linking energy security with climate change.

India's Climate Change Identity in 2012

The MCA of the 2012 reportage, a month before and after the COP-18 held in Doha, showed a slightly different trend to that of 2009 reportage (Table 3.9).

Out of the 100 articles analysed, the *entrepreneurial identity* received maximum attention in the media reporting on climate change in 2012 with 36% sample texts linking innovation with the climate change discourse. A 6% rise was noticed as compared to 2009 reportage (30%). Similar number of articles showed ambiguities in both 2009 and 2012 (1%), and 63% of the articles ignored the *entrepreneurial identity*.

The MCA showed that 32% of the articles were affirmative regarding India's *emerging nation identity* with respect to climate change negotiations, and 1% dip in reportage was noticed from 2009 reportage.

Continuing the trend of declining reportage, the *developing nation identity* also experienced a downturn from 29% in 2009 to 27% in 2012. Ambiguities in the articles almost doubled from 2009 reportage with 12% of the sample texts portraying uncertainty in placing India under the poor country label.

Out of the 100 sample texts, 25% of the articles reported on the *industrial identity*, discussing industries and industrial nations responsible for carbon emissions in the atmosphere. Similar to the other identities, *industrial identity* experienced a drop in the reportage by 2% as compared to the 2009 MCA.

Table 3.9 Media content analysis for 2012 reportage on climate change (Doha 2012)

Identities	Yes (%)	No (%)	Ambiguous (%)	Not mentioned (%)
Entrepreneurial	36.0	0.0	1.0	63.0
Emerging nation	32.0	0.0	3.0	65.0
Developing nation	27.0	0.0	12.0	61.0
Industrial	25.0	0.0	0.0	75.0
Rural	22.0	0.0	0.0	78.0
Energy security	11.0	0.0	2.0	87.0

About 22% of the media reports recorded coverage on the *rural identity*. The same identity gathered responses of the order of 26% in 2009 reportage.

The MCA showed that the *energy security identity* gained more credence in 2012 with 11% of the articles linking energy security with climate change negotiations (rise from 8% reportage in 2009). Only 2% of the media reports showed ambiguities. Rising energy needs due to urbanisation, dependence on coal and need for exploiting alternate forms of energy for meeting the demands of the growing population were explicitly referred to in these articles.

India's Climate Change Identity Across Five Mainstream Newspapers During COP-15, 2009

The 100 articles from the sample set of 2009 were further analysed across the five mainstream English online newspapers of India with respect to the six climate change identities. The responses of these media reports are given in Table 3.10.

Out of the 100 articles randomly selected (using an online software) for 2009 reportage on climate change, 4% belonged to *ET*, followed by 44% from *The Hindu* and 8% from *HT* and 43% of the total articles were from *TOI*. Only 1% of the articles from *HBL* for the year 2009 matched the search criteria. It is important to mention that for the ease of comparison and analysis, articles that agreed with the identities ('yes' column) were used to populate Table 3.10. Other columns calculating 'no', 'ambiguous' and 'not mentioned' were removed from this table.

Table 3.10 Media content analysis for five mainstream newspapers reporting on COP-15, 2009

Identities/newspapers	ET (%)	The Hindu (%)	HBL (%)	HT (%)	TOI (%)
Industrial	50.0	15.9	0.0	25.0	37.2
Emerging nation	100.0	27.3	0.0	37.5	32.6
Rural	25.0	25.0	0.0	37.5	25.6
Entrepreneurial	25.0	36.4	0.0	25.0	25.6
Developing nation	75.0	29.5	0.0	37.5	23.3
Energy security	0.0	13.6	0.0	12.5	2.3

Out of the media reports from *ET*, all the articles discussed India's *emerging nation identity*. While 75% of the articles mentioned the *developing nation identity*, 50% of the articles mentioned the *industrial identity*. The *rural identity* and the *entrepreneurial identity* received similar attention of the order of 25%. Surprisingly, the *energy security* aspect linked to the climate change negotiations were completely ignored from the reportage of ET in 2009.

The Hindu media reports (44 out of the total 100 sample texts for 2009) debated mostly about India's *entrepreneurial identity* (36.4%) in terms of energy conservation and efficiency measures and building towards low-carbon economy. Asserting India's *developing nation identity*, 29.5% of the articles asserted inclination more towards the G-77 assemblage at the climate summits. Followed by the *emerging nation identity* (27.3%), the articles acknowledged India's potential to be a proactive leader in shaping the climate deal. The *rural identity* scored the fourth highest (25%) in *The Hindu* reportage on climate change in 2009. Citing the *industrial identity*, 15.9% of the articles blamed industrialised nations for carbon emissions.

HT (eight out of the total of 100 sample texts for 2009) demonstrated a different trend. Reportage on India's *emerging* and *developing nation identity* was at par with 37.5% of the sample texts mentioning India's association with the G-77 and hyphenation with China. The *rural identity* bagged responses from similar number of articles (37.5%), linking poverty, rural economy and affordability narrative to climate change. The *entrepreneurial* and *industrial identity* was reported by 25% of the sample texts analysed. Scoring the lowest, the *energy security identity* was cited by only 12.5% of the articles from *HT*.

Out of 43 articles from *TOI*, the maximum share of reportage unexpectedly belonged to the *industrial identity* (37.2%). These articles at some point or the other attributed industries or industrialised nations for carbon emissions currently present in the environment. The *emerging nation identity*, hyphenating India with China or focusing on India's BASIC affiliation, was emphasised by 32.6% of the articles; and 25.6% of these articles reported on the link between climate change and the poverty-income trap, along with placing emphasis on the vulnerabilities due to climate change. The *entrepreneurial identity* was highlighted by the same number of articles (25.6%). India's inclination towards developing nations' stance at the climate summit and its membership in the G-77 grouping was reported by 23.3% of the articles. The *energy security identity* received scant attention (2.3%) in the sample texts analysed from *TOI* in 2009.

HBL recorded no assertion on any of the six identities. The articles focused on the impact of climate change and global status of the negotiations. The nature of reporting was not impact specific. It is important to mention that only one article from *HBL* was picked by the randomiser software (used to select an impartial sample set of media articles).

INDIA'S CLIMATE CHANGE IDENTITY ACROSS FIVE MAINSTREAM NEWSPAPERS DURING COP-18, 2012

Similar to 2009 analysis, the 100 articles from the sample set of 2012 were further examined across the five mainstream English online newspapers of India with respect to the climate change identities. This is shown in Table 3.11.

Out of the 100 articles analysed of the 2012 reportage on climate change, 37% belonged to *ET*, followed by 13% from *The Hindu* and 15% from *HBL*. *HT* and *TOI* contributed 12% and 23%, respectively, to the sample text for 2012.

Dissimilar to the 2009 reportage, the maximum reportage on *industrial identity* came from *ET* with 48.6% articles in 2012. However, there is a slight drop as compared to the 2009 reporting (50%). The *emerging nation identity* also experienced reduction in attention in 2012 – from 100% in 2009 to 45.9% in 2012. The *entrepreneurial* and India's *developing nation identity* were focused upon by equal number of articles (37.8%). While the *rural identity* was evident in 21.6% of the sample texts analysed, only 2.7% linked energy scarcity discourse with climate change.

Table 3.11 Media content analysis for five mainstream newspapers reporting on the COP-18, 2012

Identities/newspapers	ET (%)	The Hindu (%)	HBL (%)	HT (%)	TOI (%)
Industrial	48.6	0.0	20.0	16.7	8.7
Emerging nation	45.9	0.0	33.3	50.0	17.4
Developing nation	37.8	0.0	13.3	50.0	21.7
Entrepreneurial	37.8	38.5	46.7	41.7	21.7
Rural	21.6	23.1	26.7	8.3	26.1
Energy security	2.7	7.7	33.3	8.3	13.0

The *Hindu* showed a substantially different trend in its 2012 reportage on climate change. The *entrepreneurial identity* was evident in 38.5% of the articles. This was slightly higher than the coverage in 2009 (36.4%) and scores the highest among all the other identities mentioned by *The Hindu* in 2012. The *rural identity* noticed a fall by 1.9% in reporting as compared to 2009. The *energy security identity*, on the other hand, experienced a noteworthy decrease in the reportage from 13.6% in 2009 to 7.7% in 2012. These articles focused on increasing dependency on imports of primary and secondary forms of energy and energy scarcity in the developing nations. Surprisingly, none of the articles mentioned any aspect of *industrial, developing* or *emerging nation identity* in 2012 – starkly different from 2009 reportage, where these identities were referred by 15.9%, 29.5% and 27.3%, respectively.

While there were hardly any *HBL* media reports in 2009 MCA analysis, 15% of the 100 sample texts for 2012 were contributed by the *HBL*. Similar to *The Hindu* reportage, *HBL* articles laid more emphasis on the *entrepreneurial identity* (46.7%). The second highest scoring identity in terms of attention received was the *energy security identity* at 33.3%, followed by the *rural identity* with 26.7% responses. While the *emerging nation identity* attracted affirmative responses from 33.3% of the sample texts, India's *developing nation identity* was pushed down to the bottom of the ladder, evident in just 13.3% responses.

HT laid equal emphasis on India's *developing* and *emerging nation identity* – securing 50% of the total responses for each of these identities. The *entrepreneurial identity* scored the second highest with 41.7% of the sample texts responding to the green market and innovations related to energy efficiency. About 16.7% of the articles mentioned culpability of the industrialised nations towards rising emissions in the atmosphere. Only 8.3% of the sample texts divulged into linking climate change with energy security imperatives or with affordability narrative. This is noticeably different from the 2009 reportage, where 12.5% and 37.5% of the articles responded to *energy* and *rural identity*, respectively.

TOI articles highlighted the *rural identity* with 26.1% responses – slightly higher than the 2009 reportage (25.6%). The *entrepreneurial* and *developing nation identities* were at par with each other – 21.7% responses from the sample texts. The *emerging nation identity* experienced a dip by 15.2% from 2009 reportage with only 17.4% of the articles hyphenating India with China and considering it a part of the BASIC grouping. The *energy security identity*, on the other hand, showed an increase of 10.7%. The *industrial identity*, focusing on industrial emissions, received scant attention at 8.7% – a drastic reduction from 2009 reporting (37.2%).

Very broadly, the climate change discourse in India situates itself in the intervening space between (a) the demographic divide between the rural and urban populations and the industrial, entrepreneurial and social impulses required to bridge that gap; and (b) the hotly contested political space for the attention and patronage of the disproportionately higher rural agrarian populace. Within the bounds of this framework, India's climate change discourse alternates between various identities, depending on which aspect of the discourse is on the ascendant.

For instance, the emphasis on the *industrial* and *entrepreneurial self-identities* in the pre-COP climate discourse in the Indian media, and strong emergence of the *rural self-identity* during and after the COP, situates itself in this narrative (Figs. 3.1 and 3.2). However, what cannot be ignored is the preponderance of the *developing nation identity* during COP and the residual carry forward of this enhanced narrative post-COP.

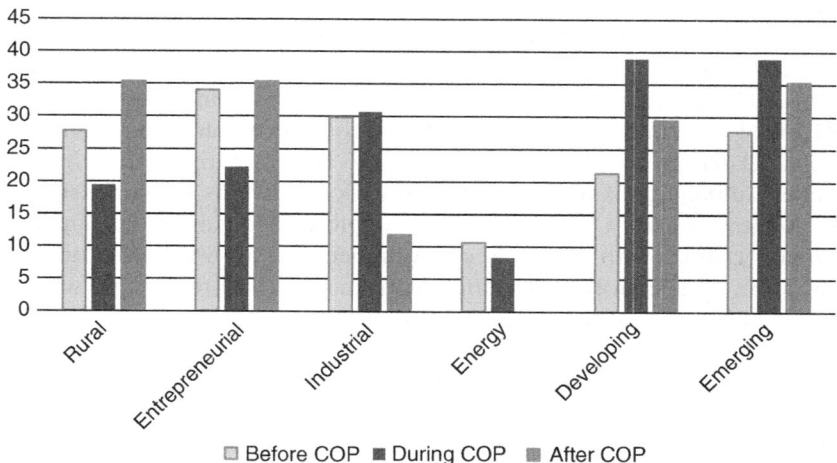

Fig. 3.1 MCA 2009: Before, during and after COP-15 reportage (percentage of all analysed media articles including each identity)

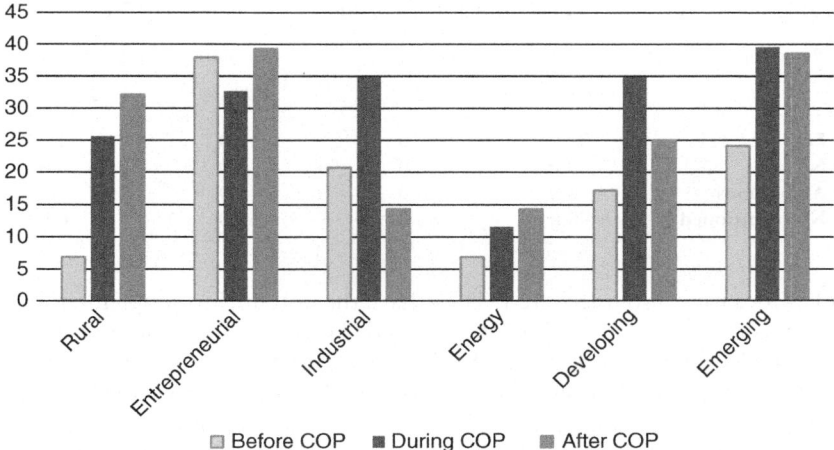

Fig. 3.2 MCA 2012: Before, during and after COP-18 reportage (percentage of all analysed media articles including each identity)

Presented alongside the scientific narrative are (a) commentary on the centrality of industrial growth for India to accomplish her development goals, and towards this end, the limited choices available for India to move away from the traditional carbon-intensive energy options to achieve this, and (b) the entrepreneurial and economic opportunity presented in the development, manufacturing and trading of carbon-friendly methodologies and technologies, referred to in this research as *industrial* and *entrepreneurial self-identities*. This aligns with previous studies on the subject (e.g. Leiserowitz and Thaker 2012; Billett 2009). Leiserowitz and Thaker (2012) posit that when left to their own means, Indian media commentary on climate change draws heavily from the scientific discourse, often closely mirroring it, with media being trusted almost at par with the scientific community themselves, which as per their study were the single most trusted composite group on the subject of climate change.

Auxiliary Findings of 2009 and 2012

While conducting the MCA for 100 sample texts from COP-15, 2009, and COP-18, 2012, certain substantial patterns outside the focus on six

Table 3.12 Auxiliary findings of COP-15, 2009 and COP-18, 2012 MCA

Responses	Mixed identity		Climate pornography	
	2009 (%)	2012 (%)	2009 (%)	2012 (%)
Yes	22.0	17.0	42.0	44.0
No	0.0	0.0	2.0	0.0
Ambiguous	0.0	0.0	1.0	0.0
Not mentioned	78.0	83.0	55.0	56.0

identities were also uncovered. These have the potential to supplement the analysis of MCA and couldn't be ignored. Hence, this section lists down the auxiliary findings from MCA of 2009 and 2012.

One of the most prominent finding was the usage of the term 'developing' and 'emerging' for India in the same article. For instance, the articles hyphenating India with China or BASIC also declared India's *developing nation identity* and its inclination towards the G-77 bloc. For further evaluation, these could be categorised as *mixed identity*.

The second perceptible response gathered from the media reports of 2009 and 2012 were directed towards the impact of climate change. Over 40% of the articles mentioned potential losses, death and destruction likely to be caused by climate change. Since these media reports arrested attention through their alarmism and sensationalism, these could be categorised as *climate pornography*.

The responses for the 'mixed identity' and 'climate pornography' for 2009 and 2012 are presented in Table 3.12. The mixed identity responses were calculated by combining the results of *developing* and *emerging nation identity* codes. For example, articles referring to both the identities were taken as affirmative for the *mixed identity*. *Climate pornography* results were directly calibrated based on reference to impacts of climate change in the media article being analysed.

CHAPTER 4

The Personal Choices of Indian Experts

Abstract This chapter outlines the approach taken by various experts in Indian climate change negotiations. This is based on a series of interviews conducted by the lead author. It will explore the use of the multiple identities by these experts.

Keywords Interviews · Expert opinions · Security · Emerging · Developing

Interviewees were selected from five main stakeholder groups that inform policies around climate change in India: government officials, media personnel, non-govermental organisations (NGOs)/think tank professionals, academics and private sector professionals. About 30% of the interviewees belonged to the NGO/think tank group, 25% to government organisations, 20% to academic institutions, 20% to the corporate sector and the remaining belonged to the media. While selecting the interviewees, gender and age balance were sought carefully. About 43.6% of the interviewees were females and 35% belonged to the younger age group—between 25 and 35 years.

© The Author(s) 2017
S. Saran, A. Jones, *India's Climate Change Identity*,
DOI 10.1007/978-3-319-46415-2_4

Rural Identity

A set of four questions were asked of 30 interviewees on India's *rural identity*. Touching upon the aspects such as employment, agricultural subsidies, demand side management and vulnerabilities, responses to the questions are given in Table 4.1.

Out of the 30, 53.3% of interviewees suggested that the agrarian economy does influence India's international approach to climate change owing to a large dependant population. Dependence of the rural population remains high both directly in terms of farm income and indirectly because of farm subsidies that impact farm incomes. Around 61.5% of the female interviewees supported this logic as compared to 47.1% males. Mid-level and young professionals also dominantly choose this response counting at 70% and 50%, respectively.

Approximately 3.3% of the interviewees highlighted Indian agriculture's vulnerability to the changing precipitation patterns; therefore,

Table 4.1 Does agriculture sector influence India's position at climate negotiations?

Responses	Total findings (%)	Female responses (%)	Male responses (%)	Mid-level professionals (%)	Young professionals (%)
Yes, because large population is dependent on subsidised farm inputs and farm income	53.3	61.5	47.1	70.0	50.0
Yes, India is vulnerable to climate change vulnerability	3.3	0.0	5.9	0.0	0.0
Yes as India must seek R&D, funding and technology	23.3	15.4	29.4	10.0	0.0
No, India is increasingly urbanising	16.7	23.1	11.8	10.0	50.0
No, emission from agriculture is low	3.3	0.0	5.9	10.0	0.0
Ambiguous	0.0	0.0	0.0	0.0	0.0
Not answered	0.0	0.0	0.0	0.0	0.0

submitted that agrarian dependence is likely to influence India's decisions at the climate summits. None of the females responded on this, but 5.9% of the males agreed to the stance.

About 23.3% of the respondents stated that India's demand for research, development, funding and technology for adaptation stems from the vulnerability of its large rural population. And 29.4% male respondents, followed by 15.4% female respondents, supported this argument. While only 10% mid-level professionals agreed to this reason, there were zero young professionals who responded to this.

On the other hand, only 16.7% of the respondents contradicted this stand and asserted that the increasing rate of urbanisation will reduce dependency on agriculture and therefore 'agrarian economy' would have less of an impact on the negotiations. About 23.1% of the females supported this view as compared to 11.8% male respondents; 50% of the young professionals and only 10% of the mid-level experts asserted that the rate of urbanisation is a factor at climate negotiations.

Approximately 3.3% of the interviewees said that the emissions from agriculture are very low as compared to other sectors and even other countries. Therefore, agrarian dependence should not be a point of focus for India at the climate summits. About 5.9% of the males and 10% of the mid-level experts responded positively to this.

Table 4.2 includes responses on the question of agricultural subsidies. Approximately 23.3% of the respondents believed the subsidies to be non-negotiable as the farm sector is heavily dependent upon external inputs. About 29.4% male and 15.5% female respondents were of this opinion, and 25% of the young professionals reaffirmed this view. Only 10% of the mid-level professionals supported this view point.

While it was generally agreed that subsidies should be slowly phased out, 26.7% interviewees said that subsidies are politically impossible to do without. Some of the respondents suggested that despite the subsidies being unsustainable for the government, vote bank politics makes this impossible. About 30.8% of the females and 23.5% male respondents proposed this explanation. Responses from mid-level and young experts were comparable at 25% and 30%, respectively.

A total of 6.7% of the interviewees were of the view that food security obligations were said to influence the agricultural subsidy policies and were the main concern for India. About 7.7% female and 5.9% male respondents averred to this. Only 10% of the mid-level experts considered

Table 4.2 Can agriculture subsidies be removed? (Yes or no and explain why)

Responses	Total findings (%)	Female responses (%)	Male responses (%)	Mid-level professionals (%)	Young professionals (%)
No, because those involved in this sector are dependent on this input	23.3	15.4	29.4	10.0	25.0
No, because they are politically non-negotiable	26.7	30.8	23.5	30.0	25.0
No, as India has food security issues	6.7	7.7	5.9	10.0	0.0
Yes, as intended consumers are left out	23.3	23.1	23.5	30.0	25.0
Yes, better to invest in rural infrastructure and agriculture research	6.7	7.7	5.9	10.0	0.0
Yes, agriculture sector must be liberalised	10.0	15.4	5.9	10.0	25.0
Others	0.0	0.0	0.0	0.0	0.0
Ambiguous	3.3	0.0	5.9	0.0	0.0
Not answered	0.0	0.0	0.0	0.0	0.0

food security as India's main internal concern. None of the young professionals felt this to be of concern.

Almost 23.3% of the respondents felt that the farmers who are most in need of subsidies are usually left out. Therefore, the benefits of the subsidies do not reach the intended customers. Males and females rejoined similarly on this with 23.5% and 23.1% responses, respectively. Similarly, 30% of the mid-level and 25% of the young experts questioned the effectiveness of the subsidies and placed it as India's main internal concern.

Rural infrastructure—farm mechanisation, improved agricultural practices, technologies and market access—is highly inadequate in India, noted 6.7% of the respondents. About 7.7% of the females and 5.9% of the males suggested that agricultural subsidies would be difficult to negotiate in the absence of support for rural infrastructure development. And 10% of the mid-level professionals also agreed that rural infrastructure development should be one of the priorities, if subsidies were to be phased out.

Only 10% of the interviewees said that lack of proper liberalisation in the agro-sector is keeping the subsidies from being phased out. About 15.4% of the female respondents suggested that liberalising agricultural trade would allow greater market access, export assistance and eventually reduce the burden on the subsidies; 5.9% of the male respondents settled on this; and 25% of the young and 10% of the mid-level professionals suggested removing barriers for agricultural trade.

Only 3.3% of the total respondents and 5.9% of the male interviewees were ambiguous in their answers to the question on subsidies in agriculture

A large proportion of respondents agreed that significant energy efficiency savings can be made through demand side management in the agricultural sector. About 23.3% respondents suggested that savings through energy efficiency should be a long-term mitigation goal (Table 4.3). Almost 35.3% of the male respondents and 7.7% of the females agreed with this. One-fifth of the mid-level experts and a fourth of the young professionals believed that efficiency in the agriculture sector can lead to emission reduction in the long term.

Agreeing with the logic of efficiency gains in the agricultural sector, 10% of the total respondents said that demand side management will reduce the burden on the subsidies and improve agricultural productivity. The response among females (15.4%) was stronger than among male counterparts (5.9%). While the young professionals did not consider this aspect important, 20% of the mid-level experts agreed with the rationale.

Equally, 10% of the respondents agreed with the logic of energy efficiency in the agriculture sector that pivots on the integrated water energy management approach. About 11.8% of the male and 7.7% of the female respondents suggested that focus on demand side management would provide an impetus to integrated energy management. Given that agriculture utilises the largest portion of subsidised energy and freshwater, 20% of the mid-level professionals proposed a focus on energy water efficiency within the demand side management paradigm. None of the young professionals saw this as a significant opportunity.

On the other hand, 26.7% of the respondents argued that emission reduction from agricultural sector through energy-efficient practices would be nominal as compared to industries. Agreeing to this logic, 29.4% of the males and 23.1% of the female respondents also emphasised the larger

Table 4.3 Is there scope for energy efficiency savings in the agricultural sector through demand side management?

Responses	Total findings (%)	Female responses (%)	Male responses (%)	Mid-level professionals (%)	Young professionals (%)
Yes, there are long-term mitigation opportunities	23.3	7.7	35.3	20.0	25.0
Yes, will reduce burden on subsidies	10.0	15.4	5.9	20.0	0.0
Yes, will help integrated water-energy management	10.0	7.7	11.8	20.0	0.0
No, only nominal contribution as compared to industries	26.7	23.1	29.4	20.0	25.0
No, as urban India's carbon footprint is much larger	13.3	23.1	5.9	10.0	25.0
No, better to focus on power generation efficiency	16.7	23.1	11.8	10.0	25.0
Others	0.0	0.0	0.0	0.0	0.0
Ambiguous	0.0	0.0	0.0	0.0	0.0
Not answered	0.0	0.0	0.0	0.0	0.0

opportunity of decreasing industrial emissions. About 20% of the mid-level and 25% of the young professionals submitted that focus on industries instead of agricultural sector would yield better results in terms of mitigating climate change.

Approximately 13.3% of the interviewees were of the view that energy efficiency savings in agriculture would not have a large impact as most emissions are urban. Female respondents (23.1%) and young professionals (25%) were more inclined towards this proposition than the male interviewees (5.9%) and the mid-level experts (10%).

Focusing on power generation instead of agriculture was suggested by 16.7% of the total interviewees. Females (23.1%) approved of bringing efficiency in power generation as a better option than focusing on agriculture, followed by only 11.8% male respondents, and 10% mid-level and 25% young professionals.

Overall, respondents were nearly equally divided on the question of whether demand side efficiency gains can be catalysed in agriculture and if this should be a matter of priority.

Table 4.4 populates responses on the question of India's adaptation response to climate-induced vulnerabilities. A large number of respondents were of the view that India should seek global partnership in mitigation, R&D and tech transfer to safeguard its large population vulnerable to the changing climate. About 53.3% of the respondents supported this proposition, of which 46.2% were females and 58.8% were male respondents; and 70% of mid-level experts and a fourth of the young professionals also supported this adaptation-centric rationale.

One of the propositions put forth to the interviewees was that with climate-related vulnerabilities India should develop innovative insurance products by allowing enhanced innovation and international participation. About 26.7% of the interviewees agreed that greater insurance innovation to hedge against climate vulnerabilities would be appropriate, representing

Table 4.4 How will India respond to climate-induced vulnerabilities in the agriculture sector?

Responses	Total findings (%)	Female responses (%)	Male responses (%)	Mid-level professionals (%)	Young professionals (%)
Seek global partnership in mitigation, R&D and tech transfer	53.3	46.2	58.8	70.0	25.0
Develop innovative insurance products by allowing FDI	26.7	46.2	11.8	20.0	75.0
Maintain CBDR principles and seek support from rich countries	10.0	0.0	17.6	10.0	0.0
Others	3.3	0.0	5.9	0.0	0.0
Ambiguous	6.7	7.7	5.9	0.0	0.0
Not answered	0.0	0.0	0.0	0.0	0.0

the views of a much larger proportion of female respondents (46.2%) than male respondents (11.8%). And 75% of the young professionals felt that this was the way forward instead of seeking financial and technological support from the developed countries, while 20% of the mid-level professionals agreed with this approach.

Maintaining position on common but differentiated responsibilities (CBDR) at the climate negotiations was supported by only 10% of the respondents. While none of the female interviewees mentioned this as an effective step towards reducing vulnerability of India's rural population, 17.6% of the males articulated that it is through the CBDR principles that India or other developing countries can demand the right to development and thereby reducing dependence on agriculture. And 10% of the mid-level professionals claimed that CBDR is an essential component of India's position at the global climate negotiations.

Energy Security Identity

India's energy security identity was explored by questioning the interviewees on aspects such as import dependency, efficiency in transmission and distribution, primary energy demands and alternate forms of energy. Main findings from 30 interviews are listed in Table 4.5.

Table 4.5 records responses on the question of development of renewable energy in India. About 20% of the respondents were positive that developing renewable energy would help reduce import bills (Table 4.5). This would make India's position at the climate summits stronger, said 15.4% of the female respondents, followed by 23.5% male interviewees. A total of 40% of the mid-level professionals readily supported development of renewables to reduce dependence on imported fuels, while none of the young professionals thought of this as a viable alternative to conventional energy.

Approximately 53.5% of the respondents were sceptical of renewables meeting primary energy demands in the short term; 38.5% of the female and 64.7% of the male respondents stressed that India will continue to rely on coal and imports to meet its developmental objectives; and 30% of mid-level and 50% of young professionals supported this position.

Lack of political will is often seen as a reason for slow development of renewables in the country. About 16.7% of the total respondents supported this reasoning, with a larger proportion being females; 23.1% of female and 11.8% of male interviewees emphasised that despite having

Table 4.5 The development of the renewable energy sector is inevitable? (Yes or no and explain why)

Responses	Total findings (%)	Female responses (%)	Male responses (%)	Mid-level professionals (%)	Young professionals (%)
Yes, renewables to help reduce import bills	20.0	15.4	23.5	40.0	0.0
No, renewable can't replace primary energy	53.3	38.5	64.7	30.0	50.0
No, lack of political will obstruct renewables	16.7	23.1	11.8	20.0	25.0
No, renewables are only driven by climate negotiations	10.0	23.1	0.0	10.0	25.0
Others	0.0	0.0	0.0	0.0	0.0
Ambiguous	0.0	0.0	0.0	0.0	0.0
Not answered	0.0	0.0	0.0	0.0	0.0

ambitious national-level plans, India's slow progress in deploying wind, solar and biomass energy is mainly due to lack of political drive and vested interests; and 20% of the mid-level and 25% of the young professionals agreed with this.

About 10% of the respondents suggested that renewable energy development in India is driven by climate negotiations and 23.1% of female interviewees supported the influence of global negotiations on climate and domestic renewable energy development policies. None of the male interviewees touched upon this suggestion. Only 10% of the mid-level experts and 25% of the young professionals drew a correlation between climate negotiations and renewable energy development in India.

Table 4.6 provides an analysis of responses from the interviewees on the issue of India's approach towards coal production and consumption. Half the respondents believed that India will continue to rely on coal in the future and should concentrate on better utilisation of domestic coal reserves. Interestingly on this occasion the share of response between male 58.8% and female 46.2% respondents was nearly equal. Domestic issues such as land acquisition, slow environmental clearance processes and

Table 4.6 Given that a large share of primary energy generation is based on coal (and India has among the world's largest reserves), how will India approach issues around coal production and consumption?

Responses	Total findings (%)	Female responses (%)	Male responses (%)	Mid-level professionals (%)	Young professionals (%)
Focus on better domestic coal utilisation	53.3	46.2	58.8	40.0	50.0
Better policy direction for the sector	13.3	15.4	11.8	0.0	50.0
Developing alternate forms of energy	10.0	15.4	5.9	30.0	0.0
Encouraging private participation in the sector	3.3	0.0	5.9	0.0	0.0
Shift towards more efficient technologies (supercritical, ultra-supercritical technologies)	16.7	23.1	11.8	30.0	0.0
Others	3.3	0.0	5.9	0.0	0.0
Ambiguous	0.0	0.0	0.0	0.0	0.0
Not answered	0.0	0.0	0.0	0.0	0.0

unsustainable mining practices were some of the reasons highlighted for coal sector inefficiencies; and 40% mid-level professionals and 50% young respondents supported enhancement of efficiency with the sector.

A total of 10% of the respondents who agreed were of the view that better policy direction in the public sector would improve efficiency of coal production; of which 15.4% were female interviewees and 5.9% were males. They argued that monopolistic culture of the coal industry should be changed and highlighted the lack of competitiveness due to public sector dominance, and 50% of the young professionals agreed with this. None of the mid-level professionals mentioned this in their answer.

A total of 10% of the interviewees recommended focusing on developing alternate forms of energy as a means to reduce the burden on coal production; 15.4% of females and 5.9% males were consonant with this view. Only 30% of the mid-level professionals agreed to off-grid renewables as a feasible substitute for coal, while none of the young professionals referred to this suggestion in their response.

Encouraging private sector participation in the sector to improve competitiveness and efficiency got mentioned by 3.3% of the total respondents, including 5.9% of the male interviewees. Only senior-level experts made this suggestion.

About 16.7% of the interviewees argued for a shift towards more efficient technologies, including supercritical and ultra-supercritical technologies. Focus on technological innovations was supported by 23.1% of female respondents, 11.8% of male respondents, 30% of the mid-level professionals and none of the young professionals.

Table 4.7 records responses on prioritisation of hydro and nuclear energy within India's energy mix: 23.3% of the respondents, including 30.8% of the females and 17.6% of the males, agreed that nuclear and hydro should be prioritised to meet the growing energy demands, while 20% of the mid-level experts and 50% of the young professionals suggested that nuclear and hydro are among the safest options for clean energy.

Table 4.7 Should hydro and nuclear energy be prioritised given extant resource constraints?

Responses	Total findings (%)	Female responses (%)	Male responses (%)	Mid-level professionals (%)	Young professionals (%)
Yes, nuclear and hydroelectric energy should be prioritised	23.3	30.8	17.6	20.0	50.0
Yes, there must be stronger support for nuclear power	13.3	7.7	17.6	10.0	0.0
No, socio-ecological costs are high for hydropower	20.0	30.8	11.8	40.0	25.0
No, on account of safety issues	10.0	15.4	5.9	10.0	0.0
Needs further consideration	30.0	7.7	47.1	20.0	25.0
Others	3.3	7.7	0.0	0.0	0.0
Ambiguous	0.0	0.0	0.0	0.0	0.0
Not answered	0.0	0.0	0.0	0.0	0.0

Only 13.3% expressed support for nuclear energy. They contended that it is the most environment friendly of all energy resources and has high electricity generation potential. More male respondents (17.6%) than female respondents (7.7%) supported this claim. Only 10% of the mid-level professionals offered this rationale while none of the young professionals did.

A total of 20% of the respondents were of the view that rehabilitation and resettlement, land submergence and other environmental issues associated with hydropower development, are complex. This makes hydropower a difficult energy option. Females (30.8%) were more apprehensive regarding development of hydropower as against males (11.8%); 40% of the mid-level and 25% of the young professionals expressed serious concern for socio-ecological challenges in the hydropower sector.

Similarly, safety and accountability issues were highlighted by 10% of the total respondents including 15.4% of the females and 5.9% of the males. The Fukushima incident was specifically mentioned as a reference point by 10% of the mid-level experts but by none of the young professionals.

Uncertainty and lack of detailed assessment studies on both the energy sources compelled 30% of the respondents to suggest that there is a need for further careful evaluation. While the male respondents, accounting to 47.1%, demanded detailed study on nuclear and hydro, only 7.7% of the females concurred that both these technologies come with equally difficult challenges; and 20% mid-level and 25% young professionals agreed with this proposition.

INDUSTRIAL IDENTITY

India's *industrial identity* is referred to in terms of emissions from key sectors of the economy such as energy, transport, infrastructure and shipping. These characteristics were investigated in detail through semi-structured questions. Findings of the questions are given in Table 4.8.

Table 4.8 aggregates responses on what India should do for sustainable, inclusive growth and employment generation. About 13.3% of the respondents suggested promotion of less capital-intensive industries; 15.4% of the females and 11.89% of the males agreed to this approach. Meanwhile 50% of the young professionals were very optimistic regarding innovative start-ups to spur growth and generate employment.

One of the dominant views put forth was that development of manufacturing industries should be carefully planned keeping in view the scarcity

Table 4.8 India's growth over the last two decades has been driven by the service sector. However, for sustainable, inclusive growth, India must generate employment through development of industry (such as manufacturing). How will India manage this vital imperative?

Responses	Total findings (%)	Female responses (%)	Male responses (%)	Mid-level professionals (%)	Young professionals (%)
Promote less capital-intensive industries	13.3	15.4	11.89	0.0	50.0
Integrate planning with energy security focus	30.0	38.5	23.5	40.0	25.0
Maintain right to develop and grow in global forums	6.7	7.7	5.9	10.0	0.0
Build manufacturing sector and green economy	43.3	30.8	52.9	40.0	25.0
Others	3.3	7.7	0.0	0.0	0.0
Ambiguous	0.0	0.0	0.0	0.0	0.0
Not answered	3.3	0.0	5.9	0.0	0.0

of the resources. Integrating development of industries with energy security focus should be prioritised, according to 30% of the respondents. And 38.5% of the female interviewees and 23.5% of the male interviewees agreed and said that energy scarcity can be a growth inhibitor unless industrialisation incorporates sustainable energy management principles. While a fourth of the young professionals agreed with this view, none of the experts did.

About 6.7% of the respondents recommended that India needs to maintain its right to develop and grow at international forums. Of this 7.7% of female respondents, 5.9% of male respondents and 10% mid-level experts agreed that India's strategic position at the climate summits should take into account aspirations for double-digit growth and employment generation objectives.

Many interviewees believed that building manufacturing sector and green economy would be critical to achieve sustainable, inclusive growth and generate employment. About 43.3% of the respondents recognised this opportunity, including 52.9% of the male and 30.8% of the female interviewees. Mid-level experts (40%) and young professionals (25%) were optimistic about green economy as a way forward.

Entrepreneurial Identity

India's *entrepreneurial identity* is related to features such as green market, manufacturing sector and industrial efficiency. These characteristics were investigated in detail through semi-structured questions. Findings of the interview are mapped in Table 4.9.

About 36.7% of the respondents agreed with the proposition that India is positioning itself to be a large market for green energy (Table 4.9). The scope of clean technologies and energy efficiency is an impetus for green market in India, said 23.1% of female and 47.1% of male respondents. Approximately 20% and 25% of the mid-level experts and young professionals, respectively, suggested that India should demand that the Green Climate Fund be used for purchasing intellectual property rights for clean technology.

However, 13.3% of the respondents were sceptical and posited that India is a consumer of clean technologies and not a producer yet. A total of 15.4% of the females argued that India will continue to rely on imported technology especially for solar in the coming decade, a position held by 11.8% of the male respondents and 10% of the mid-level professionals.

Table 4.9 Is India positioning itself to be a large consumer and producer of green energy and clean technologies? (Yes or no and explain)

Responses	Total findings (%)	Female responses (%)	Male responses (%)	Mid-level professionals (%)	Young professionals (%)
Yes, India is a large market for green energy	36.7	23.1	47.1	20.0	25.0
No, India is a consumer of technology not a manufacturer (yet)	13.3	15.4	11.8	10.0	0.0
Yes, green growth is critical for climate change adaptation	33.3	53.8	17.6	50.0	75.0
Yes, must focus on manufacturing to reduce outflow of foreign exchange	10.0	7.7	11.8	10.0	0.0
Ambiguous	0.0	0.0	0.0	0.0	0.0
Not answered	6.7	0.0	11.8	10.0	0.0

Development of green market is imperative for India to deal with the changing climate, claimed 33.3% of the interviewees. Female respondents (53.8%) were more optimistic on the climate adaptation rationale for building a green economy than the male respondents (17.6%). Similarly, young professionals (75%) and mid-level experts (50%) agreed to this logic.

A total of 10% of the respondents stressed the urgency of focussing on developing the manufacturing sector to reduce the outflow of foreign exchange. About 7.7% of females and 11.8% of males reasoned similarly. Only 10% of the mid-level professionals supported this suggestion and none of the young professionals shared this view.

About 6.7% of the total respondents, including 11.8% of the males and 10% of the mid-level professionals, did not answer this question.

Most respondents agreed that rapid development of industrial efficiency cannot be envisioned without strong policy interventions at the federal level (46.7%), including 69.2% of the female and 29.4% of the male interviewees. Agility and not bureaucracy might become central to addressing industrial efficiency practices. However, it is hard to imagine a neutral state; therefore, industrial efficiency will need to be supported through other means. Approximately 75% of the young professionals and 60% of the mid-level professionals agreed (Table 4.10).

Only 23.3% of the interviewees thought that industrial efficiency can be achieved without support from the government, including 7.7% female and 35.3% male respondents. While none of the young professionals touched upon this, only 20% of the mid-level experts could imagine achievement of moderate levels of industrial efficiency without state support.

Table 4.10 Can rapid development of industrial efficiency be envisioned without strong policy interventions at the federal level?

Responses	Total findings (%)	Female responses (%)	Male responses (%)	Mid-level professionals (%)	Young professionals (%)
Yes	23.3	7.7	35.3	20.0	0.0
No	46.7	69.2	29.4	60.0	75.0
Ambiguous	3.3	0.0	5.9	0.0	0.0
Not answered	26.7	23.1	29.4	20.0	25.0

About 3.3% of the respondents were ambiguous in their claims. However, 26.7% of the total respondents did not offer any answer, including 23.1% females and 29.4% males together with 20% mid-level experts and 25% young professionals. Since this question posed a nuanced proposition, uncertainty in answering is not unexpected. The fact that a fifth of mid-level experts could not venture answers also indicates the complexity of the Indian political economy.

To further explore the *industrial identity* of India, interviewees were asked to suggest drivers of industrial efficiency (Table 4.11). About 20% of the respondents selected incentives and penalties, transfer of eco-friendly technologies from the West and energy pricing as the main drivers. Responses across gender and level of expertise differed.

Incentives and penalties were the main drivers for industrial efficiency suggested by 30.8% of the females and 11.8% males; and 40% of the mid-level experts selected this option as opposed to only 25% young professionals.

Similarly, technology transfer as a driver was seen as a distinct possibility by 23.1% of the female respondents and 17.6% of the male respondents. Surprisingly, none of the mid-level experts identified technology as a driver while 75% of the youngsters were confident that with transfer of improved technologies industrial efficiency can be leveraged.

Table 4.11 What are the drivers of industrial efficiency in the Indian economy?

Responses	Total findings (%)	Female responses (%)	Male responses (%)	Mid-level professionals (%)	Young professionals (%)
Incentives and penalties	20.0	30.8	11.8	40.0	25.0
Technology transfer	20.0	23.1	17.6	0.0	75.0
Open economy and trade	16.7	15.4	17.6	20.0	0.0
Infrastructure	6.7	15.4	0.0	20.0	0.0
Energy pricing	20.0	15.4	23.5	10.0	0.0
Optimal regulatory environment	10.0	0.0	17.6	10.0	0.0
Ambiguous	0.0	0.0	0.0	0.0	0.0
Not answered	6.7	0.0	11.8	0.0	0.0

Energy pricing was identified as a driver by primarily male respondents (23.5%), rather than female respondents (15.4%). However, only 10% of the mid-level professionals and none of the young professionals suggested this option.

Enhancing the openness of the Indian economy to improve industrial efficiency was seen to be a driver by 16.7% of the respondents, including 15.4% females, 17.6% males and 20% mid-level experts. Young professionals did not mention this as a driver for industrial efficiency.

Optimal regulatory environment to support industrial efficiency was identified by 10% of the respondents as a way forward and was mentioned only by male interviewees (17.6%) and mid-level professionals (10%).

Improved infrastructure, as a driver for industrial efficiency, received scant attention (6.7%). Only 15.4% of the females and 20% of the mid-level professionals mentioned it in their response.

Developing Nation Identity

India's *developing nation identity* highlighted by the membership of G-77 group, adhering to principles of CBDR, was explored through interviews. The main findings are listed in Table 4.12.

Interviewees were divided on the question of India continuing to be a part of G-77 over the next decade (Table 4.12). About 40% of the respondents agreed with this positioning while 36.7% did not find the platform effective. Participation in the G-77 was supported by 38.5% of the female respondents, while 30.8% emphasised the latter. Equal numbers of males (41.2%) argued for and against India's membership in G-77. Mid-level professionals (40%) and young professionals associated (75%) were more inclined towards India's alliance with the developing countries. Comparatively, only 10% and 25% mid-level and young professionals, respectively, suggested that G-77 cannot act as pressure group, taking into account fundamentally varied conditions of small island nations, low-income countries and the middle-income economies.

About 6.7% of the respondents showed concern with India's association with the G-77; 7.7% of females and 5.9% of the males expressed that G-77 grouping might obscure domestic focus and its unique requirements and developmental pathways. Only 10% of the mid-level experts responded to this. It is important to note that 13.3% of the interviewees did not answer this question, including 23.1% females and 5.9% males, and 40% of the mid-level experts.

Table 4.12 Will being a part of G-77 be beneficial to India in climate change negotiations over the next decade?

Responses	Total findings (%)	Female responses (%)	Male responses (%)	Mid-level professionals (%)	Young professionals (%)
India should remain open to G-77 and maintain CBDR principles	40.0	38.5	41.2	40.0	75.0
G-77 cannot act as a pressure group given the divergence of interests within	36.7	30.8	41.2	10.0	25.0
G-77 grouping obscures national priorities	6.7	7.7	5.9	10.0	0.0
Ambiguous	3.3	0.0	5.9	0.0	0.0
Not answered	13.3	23.1	5.9	40.0	0.0

When asked how important South–South cooperation is within the context of global leadership at climate summits, 56.7% were found to be strongly in support. Responses included 53.8% of females and 58.8% males (Table 4.13). A 75% of the young professionals considered the cooperative paradigm of South–South cooperation very important, particularly for sharing best practices and evolving joint positions, as compared to 40% of the mid-level experts.

About 23.3% of the interviewees felt that South–South cooperation is important but not essential. This view was supported more by female interviewees (38.5%) than males (11.8%). Similarly, 40% of mid-level experts were confident of the value of South–South cooperation and much fewer of the junior professionals (25%).

Only 3.3% of the interviewees discarded the importance of South–South cooperation and stated that it will not matter much at the global negotiations given the lack of coherence in interests of the member countries.

None of the respondents vetoed the very concept of South–South cooperation but 13.3% of respondents were ambiguous in their comments. This included 17.6% men and 7.7% women, and 20% mid-level

Table 4.13 How important is the South–South cooperation within the context of global leadership in climate change mitigation?

Responses	Total findings (%)	Female responses (%)	Male responses (%)	Mid-level professionals (%)	Young professionals (%)
Very important	56.7	53.8	58.8	40.0	75.0
Important	23.3	38.5	11.8	40.0	25.0
Does not matter	3.3	0.0	5.9	0.0	0.0
Not important	0.0	0.0	0.0	0.0	0.0
Ambiguous	13.3	7.7	17.6	20.0	0.0
Not answered	3.3	0.0	5.9	0.0	0.0

professionals; 3.3% of the respondents did not respond to the interview question at all.

In wake of the previous question, interviewees were asked if India's alliance with G-77 is a contradiction in itself owing to the fact that India is a trillion-dollar economy (Table 4.14). While 60% of the respondents defied the contradiction, 3.3% of the interviewees proclaimed that fissures among G-77 are already visible because of the differing focus and aspirations from the climate deal. Some of India's mitigation issues related to energy and large industries are similar to developed countries.

Table 4.14 India is a trillion-dollar economy and is significantly larger than most economies that are part of the G-77. Is this a contradiction in itself?

Responses	Total findings (%)	Female responses (%)	Male responses (%)	Mid-level professionals (%)	Young professionals (%)
Yes	3.3	0.0	5.9	0.0	0.0
No	60.0	69.2	52.9	60.0	75.0
Both	13.3	7.7	17.6	0.0	0.0
Ambiguous	6.7	7.7	5.9	10.0	25.0
Not answered	16.7	15.4	17.6	30.0	0.0

About 69.2% of the women and 52.9% of the men had no disagreement over India's G-77 membership, despite India's comparatively larger economy. Respondents explained that India shares many similarities with G-77 in terms of vulnerability and resource dependence. Certainly, from an 'adaptation' of communities' perspective, G-77 is a logical group for India to be a part of. And 60% of the mid-level experts and 75% of the youngsters supported this argument.

Notwithstanding, 13.3% of the respondents held a mixed view, considering India's membership a contradiction in places, yet logical in others—including 7.7% females and 17.6% males. India is big in quantity (aggregate as big as the G 20 group) but poor in quality (per person—as poor as G-77 countries). The contradiction is within—between quantity and quality. A total of 6.7% of the respondents were ambiguous in their arguments (7.7% of the females and 5.9% of the males; 10% mid-level professionals and 25% juniors).

About 16.7% of the respondents did not respond to the query if India's membership to G-77 is contradictory. This group included 15.4% females, 17.6% males and 30% of the mid-level professionals.

Table 4.15 shows that a majority of the interviewees (83.3%) concluded that India will remain adamant on demanding higher emission reduction from the developed nations on account of historical responsibility. About 92.3% of the females and 76.5% of the males submitted that CBDR and the right to development, defining elements of the Kyoto Protocol, will remain as cornerstones of Indian negotiating policy. While the responses from the young professionals were absolutely affirmative (100%), 80% of the mid-level professionals supported this argument.

Table 4.15 Will the principles of common but differentiated responsibility and equity continue to be the cornerstones of Indian negotiating policy?

Responses	Total findings (%)	Female responses (%)	Male responses (%)	Mid-level professionals (%)	Young professionals (%)
Yes	83.3	92.3	76.5	80.0	100.0
No	3.3	0.0	5.9	0.0	0.0
Ambiguous	3.3	0.0	5.9	0.0	0.0
Not answered	10.0	7.7	11.8	20.0	0.0

Only 3.3% of the respondents differed and stated that the changing power centres of the world and rise of emerging economies like China warrant a critical reassessment of India's negotiating policies towards the new climate deal.

Only 3.3% of the respondents were ambiguous in their replies; 10% of the total respondents did not answer the question (7.7% females and 11.8% males, and 20% mid-level professionals).

EMERGING NATION IDENTITY

India's *emerging nation identity* was explored for potential of leadership, coordinating interests with Brazil, South Africa, India and China Brazil, South Africa, India and China (BASIC)/BRICS (Brazil, Russia, India, China and South Africa) and China, and binding to legal commitments. The responses to questions pertaining to the *emerging nation identity* are given later.

Table 4.16 highlights that India's leadership role in climate summits was discussed by 50% of the respondents; and 53.8% female respondents and 47.1% males suggested that India take proactive steps in positioning itself as a leader in voicing concerns of the developing and emerging

Table 4.16 Is it incumbent upon India to position itself as a leader in climate change negotiations or is it strategically more logical for it to manage its interest through closer coordination with other developing countries?

Responses	Total findings (%)	Female responses (%)	Male responses (%)	Mid-level professionals (%)	Young professionals (%)
India should position itself as a leader in climate change negotiations	50.0	53.8	47.1	50.0	75.0
India should manage its interest through closer coordination with other developing countries	36.7	46.2	29.4	50.0	25.0
Both	10.0	0.0	17.6	0.0	0.0
Ambiguous	3.3	0.0	5.9	0.0	0.0
Not answered	0.0	0.0	0.0	0.0	0.0

nations. About 75% of the young and 25% mid-level professionals agreed with this proposition.

On the other hand, 36.7% of the respondents felt that India should align its interest through closer coordination with other developing countries; 46.2% females and 29.4% males suggested that till the new climate deal is formulated in 2015, India should maintain the principles of CBDR through close coordination rather than attempting to go the distance alone. About 50% of the mid-level professionals and 25% of the junior professionals aligned with this position.

However, 10% of the respondents said that India should strategically place itself in the middle, that is, to manage its interests with other developing as well as developed countries. Only 17.6% of the male respondents agreed to this claim, and 3.3% of the total respondents were ambiguous in their answers.

According to 26.7% of the respondents, including 7.7% females and 41.2% males, India should position itself at climate summits based on domestic needs (Table 4.17). About 20% of the mid-level professionals suggested that poverty eradication, energy security, need for capacity building and skill development will provide an impetus for India to take proactive step towards reaching a global agreement. Perhaps a surprise was that none of the young professionals alluded to any of these domestic drivers in the context of India's international bargaining position.

Table 4.17 How must India position itself at climate change negotiations?

Responses	Total findings (%)	Female responses (%)	Male responses (%)	Mid-level professionals (%)	Young professionals (%)
Protect its core domestic interests (poverty, energy, development)	26.7	7.7	41.2	20.0	0.0
Influenced by domestic political debates	20.0	23.1	17.6	20.0	0.0
Aligned to its international ambitions	50.0	69.2	35.3	60.0	100.0
Ambiguous	3.3	0.0	5.9	0.0	0.0
Not answered	0.0	0.0	0.0	0.0	0.0

About 20% of the respondents suggested that India's positioning at the international forums will be dependent on domestic political debates. Referring to the 2009 heated responses for Jairam Ramesh's speech at the climate summit, 23.1% of the females and 17.6% of the males and 20% of the mid-level professionals reflected on the significance and influence of partisan politics and national debates on foreign affairs with respect to climate change. According to these respondents, India cannot be seen as taking a proactive leadership role unless it can convince its domestic constituency.

However, the highest number of responses alluded towards India's leadership potential in consonance with its global ambitions; 50% of the total respondents including 69.2% women and 35.3% men discussed India's strong global objectives related to climate change, as somewhat disconnected from domestic needs and national debates. While 60% of the mid-level experts offered similar reasoning, acceptance for this rationale was unanimous among young professional respondents.

Table 4.18 shows that respondents were of the view that India is most likely to coordinate its interests with China/BASIC for negotiating a new climate deal. This is the likelihood predicted by 76.7% of the interviewees. Meanwhile the difference in opinions across gender and career experience categories was slimmer than in most other questions with 69.2% females, 82.4% males, 60% mid-level and 75% of the young professionals supporting this outcome.

Only 6.7% of the interviewees showed scepticism over India's inclination towards China/BASIC and proposed that India should align more

Table 4.18 Will India continue to coordinate interests with China/BASIC within the context of negotiating a new climate change agreement applicable from 2020?

Responses	Total findings (%)	Female responses (%)	Male responses (%)	Mid-level professionals (%)	Young professionals (%)
Likely	76.7	69.2	82.4	60.0	75.0
Unlikely	6.7	7.7	5.9	10.0	0.0
Ambiguous	0.0	0.0	0.0	0.0	0.0
Not answered	16.7	23.1	11.8	30.0	25.0%

Table 4.19 Can/will India bind itself to legal commitments?

Responses	Total findings (%)	Female responses (%)	Male responses (%)	Mid-level professionals (%)	Young professionals (%)
Likely	20.0	7.7	29.4	11.1	0.0
Unlikely	70.0	84.6	58.8	88.9	100.0
Ambiguous	6.7	0.0	11.8	0.0	0.0
Not answered	3.3	7.7	0.0	11.1	0.0

closely with developing nations. About 7.7% of the females and 5.9% of the males along with 10% mid-level professionals stood by this argument.

Approximately 16.7% of the total respondents did not answer this question; 23.1% of the women, 11.8% of the men, 30% of mid-level experts and 25% junior professionals provided no responses.

India is unlikely to bind itself to a legal commitment, according to 70% of the respondents (Table 4.19). Female respondents (84.6%) were more confident than male respondents (58.8%) that India will not sign a legal pledge. All young professionals thought of this as unlikely as did most mid-level experts (88.9%).

About 20% of the interviewees highlighted that India will eventually have to give some legally binding reduction commitments to be taken seriously as a responsible stakeholder in climate change negotiations. Responses from men were of the order of 29.4%—higher than the female responses at 7.7% and mid-level professionals at 11.1%.

A 6.7% of the respondents, including 11.8% males were ambiguous in their responses; and 3.3% of the total interviewees choose not to answer this question.

TRENDS IN EXPERT OPINIONS

It is evident that half of the respondents argued that India's large rural population does influence India's negotiating position at climate summits. The government sector, media and the private sector showed similarity in response (50% consensus in each stakeholder group) when grading the importance of India's large rural, agrarian population to its position at climate change negotiations. The entirety of individuals in the academic field felt this was true, but only 41.7% of the NGOs agreed with it.

India has a large agrarian population; and therefore, agricultural subsidies cannot be removed because those involved in this sector depend heavily on it. About 37.5% of the government, 50% of the media, 8.3% of the NGOs and 50% of the academicians agreed that production of the Indian farm sector is dependent on subsidised external inputs such as fertilisers, energy and so on. None of the respondents from the private sector commented on this.

A total of 50% of the media group argued that agricultural subsidies are politically non-negotiable. The statement was supported by an equal number of interviewees from the government, NGOs/think tanks, academia and private sector (25%). Focusing on energy efficiency in agricultural practices will reduce burden on subsidies, opined 25% of the government and private sector respondents. None of the other group of interviewees mentioned this in their responses.

In order to reduce climate-induced vulnerabilities, 100% respondents from media suggested that India should seek global partnership in mitigation, R&D and technology transfer. About 62.5% of the government interviewees, 58.3% of the respondents belonging to NGOs/think tanks and 50% of private sector stakeholders held consonant opinions. However, the academics interviewed were of the opinion that reducing vulnerability to climate change is a local process; innovative funding models or insurance tools would be more pragmatic than relying on multilateral funding and technology arrangements.

The government respondents with 62.5% of the responses and all other stakeholder groups (media, NGO/think tank, academia and private sector) with 50% responses maintained that renewables cannot replace primary energy demands in India.

According to 75% of the government officials and 25% of academics, India should concentrate on improving its utilisation of domestic coal. There were an equal number of consonant responses from media, NGO/think tanks and private sector stakeholders (50%).

On the question of development of nuclear and hydropower, 37.5% of the government respondents and 41.7% of the NGO/think tank stakeholders supported prioritisation of nuclear and hydropower development in the country. While 25% of the private sector interviewees agreed, none from the media and academic institutions responded to this.

In order to generate employment and achieve sustainable inclusive growth, 75% of the respondents from private sector suggested focussing on energy security; 50% of academics and 16.7% of NGOs/think tank respondents held the same view, while none from the media was of this opinion.

India is positioning itself to be a large market for green energy, claimed 50% of the three major stakeholder groups—government, media and private sector. However, only 25% of NGO/think tanks and academics agreed with this.

To achieve rapid development of industrial efficiency, strong policy interventions will be required at the federal level. About 75% of the private sector respondents agreed, along with 58.3% from the NGOs, 37.5% from the government and 25% from the academic group of interviewees.

One of the main drivers of industrial efficiency listed by the respondents was incentives and penalties. While media and private sector were silent on incentives and penalty structures for promoting industrial efficiency, 33.3% of NGO/think tanks, 25% of academics and 12.5% of the government officials supported this choice.

About 12.5% of the government officials and 50% of the academics were of the view that G-77 cannot act as a pressure group in climate negotiations due to the divergences within; 50% of the private sector and 100% of the media respondents agreed, and only 33.3% from the NGO/think tank group supported this argument.

All academics highlighted the significance of South–South cooperation in reaching a new climate deal. Of the total interviewees, 75% of the government officials, 50% of media and 41.7% of the NGO/think tank stakeholders supported this rationale as well.

All media stakeholders and academics opined that CBDR will continue to be the cornerstone of Indian negotiating policy; 87.5% of government respondents and 83.3% of NGO/think tank stakeholders supported this argument, along with 50% from the private sector.

About 37.5% of the respondents from the government and 75% from NGO/think tanks were of the view that India should position itself as a leader in climate negotiations. While respondents belonging to the media did not have this opinion, 25% of the interviewees from academia and 50% from the private sector supported India's leadership potential at climate summits.

A total of 75% of the interviewees belonging to academia felt that India will justify leadership at Conference of Parties (COP) in accordance with its international leadership ambitions; 62.5% of government officials, 50% of NGO/think tanks and 25% of private sector respondents concurred to this position. However, media provided no response to the linkage between India's global aspirations and its leadership at COP.

All the respondents from media and academia agreed that India is most likely to coordinate interests with China/BASIC; and 87.5% government and 75% of the private sector respondents were in consonance with them, whereas there was some divergence in the views of NGO/think tank representatives (58.3%).

All the respondents from academia asserted that India will not bind itself to legal commitments; 75% of the government officials, 66.7% NGO/think tank representatives and 50% of media and private sector agreed.

Before proceeding with more specific analysis, it may be useful to discuss Fig. 4.1 to understand in a broad sweep what the climate intellectuals, activists and policy elite voice on the subject.

The above expert propositions in many ways are symptomatic of the multiple strands of responses (flowing from individual and collective 'lived experiences'), argumentation and rhetoric that dominate the Indian discourse, and no doubt to understand and decipher this conversation becomes problematic for many within India and certainly for any overseas observer.

Within the same pool of interviewees (and many a times in the responses of the same individual) there were inherent contradictions and counter-intuitive reasoning on some vital issues that affect both the Indian policymaking landscape and certainly its global positioning. For instance, a strong support for India's South–South (G-77) cooperation (collaboration) agenda is observed. An even stronger support is voiced for its role in the BASIC and with China, implicating the *emerging nation identity*, and alongside there is a discernable desire for India to take leadership at global forum and on climate change issues generally.

We also observe strong support for the growing renewables energy opportunities and green growth, and at the same time there is unequivocal positioning of coal as being central to India's energy needs, and the need to focus on better utilisation of this resource. And finally just for illustrating the counter-narratives shaping the multitude in the unity, we uncover a desire for global leadership dramatically tempered by the rural narrative which implicates a continuing need to subsidise energy, water and other agriculture inputs for the rural poor.

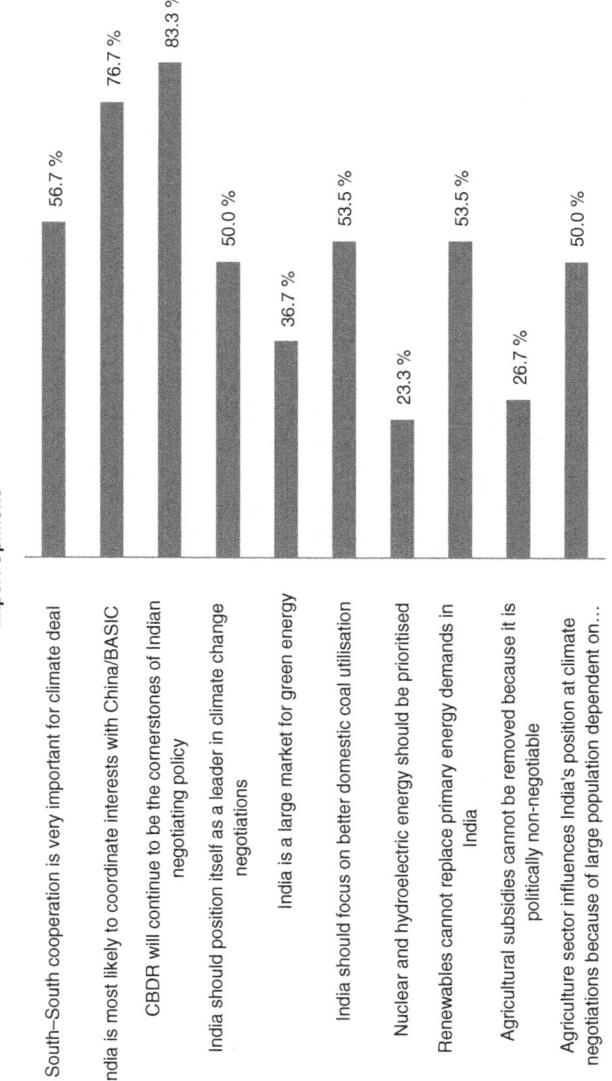

Fig. 4.1 Expert opinions (analysis of the interviews)

For the political elite, taking any position that could be interpreted as impacting an already fragile energy dynamic is viewed as being fraught with danger of erosion of political capital. Trumbo and Shanahan (2000) observe that climate communication happens through localised frameworks, and, in open societies, because of their capacity to influence public opinion and voting patterns, is closely linked to political response and the formulation of public policy. Billett (2009) argues that because of the politically sensitive nature of the discourse, political response by India to climate change has remained defensive.

Climate Colonialism

India's positioning also builds on the evocation of the colonial argument that the developed North, despite their high-carbon footprint and historical responsibility, are forcing their agenda on the less developed South, who are struggling to get their masses out of poverty, colouring the discourse in the colonialist hue of the West dominating the rest. In her impassioned speech at the international climate change negotiations in Durban, South Africa, the then Indian Environment Minister, Jayanthi Natarajan, posited, 'Does climate change mean you give up equity? India will never be intimidated by threats or any kind of pressure like this' (Natarajan 2011). This position is substantiated by expert preferences in India, which saw 83.3% supporting CBDR as the cornerstone of negotiating policy.

This strong position, espoused by Minister Natarajan in 2011, may draw inspiration from the statement on the subject by Indian Prime Minister Indira Gandhi in 1972 (Gandhi 2008). When the latter prioritised poverty and development over environmental action as suggested by developed countries, the former captured the same in her invocation of the term 'Equity'. This term did not resonate with India alone; it was a concept that was negotiated and agreed to by parties at the Earth Summit in 1992 (United Nations Environment Programme 1992). This itself can be argued to have emerged from a notion of solidarity and empathy that brought together former colonies and allowed them to negotiate effectively with some of their former colonisers.

Some climate commentators argue that there is a 'normative assumption' in India that the North's neo-colonialist desire is reflected in their aggressive articulation of a restrictive climate policy, which is aimed to stymie a rising India, a large part of whose development efforts are contingent upon the supply of affordable energy from carbon-unfriendly coal (Billett 2009; Radcliffe 2005). This assumption, they (ibid.) argue, fuels a nationalistic response reinforcing the non-acceptance of binding emission targets.

During the interviews, experts exerted that India's primary energy needs cannot be met by renewables only (Fig. 4.2). In the short term and given the pressing need of providing energy access to all, coal will be the mainstay of the Indian economy.

Within this discourse of emotional evisceration, the cause and effect are separated with recognition that while on the one hand climate change is having devastating effects on India, the cause of it lies outside India, with the developed world, and in implication, the response must reside outside. However, McManus (2000) argues that this is an important strand of the climate discourse in the developing South, where cause and effect get disaggregated along the North–South fault lines, especially so in India where such a discourse draws attention away from the inherent socio-economic inequity that exists within India (McManus 2000; Lankala 2006).

New Opportunities

The markup of political content in the energy debate witnessed through and post the COP, articulated as energy politics, however, did not diminish the extant discourse on energy opportunity or the *entrepreneurial self-identity*. It may be argued that this insularity and constancy of *entrepreneurial self-identity*, removed from the vicissitudes of the political expediency inspired *rural self-identity*, arises from the close inter-linkages between politics and business. Building political capital and capturing votebank imagination requires tremendous financial capacity. Like elsewhere, resource mobilisation in a large part is through legitimate donations by businesses. This acts as a strong incentive for the political elite to not disrupt commercial and enterprise activity. Business, in turn, is based on the matrix of profit and loss, answerable only to a limited set of stakeholders, and subject to the forces of the market, and not the civil discourse or scientific narrative on climate change.

Businesses' interest in climate change therefore is restricted to the economic opportunity presented by climate change, by way of a new market for developing and selling carbon-friendly technologies and

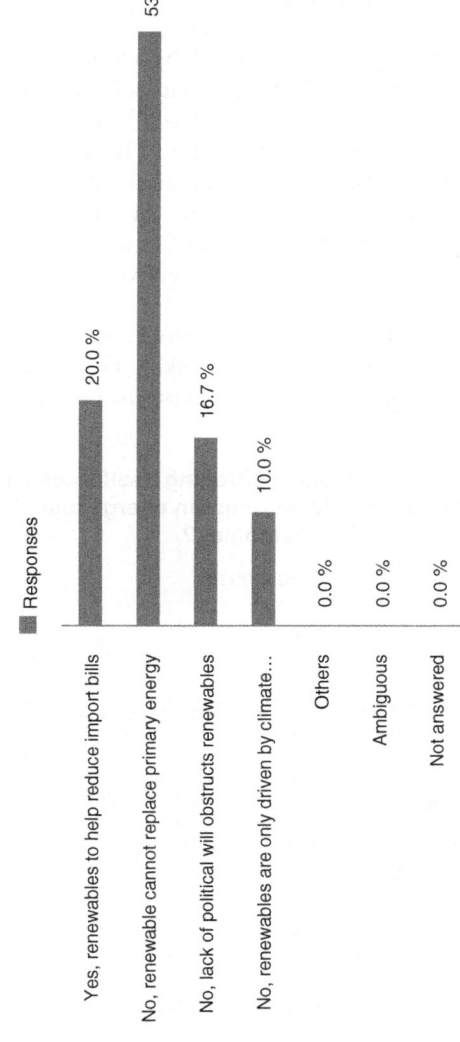

Fig. 4.2 Interview analysis: India's primary fuel demand and climate change

mechanisms, and maximising their stakeholder value. Climate experts agree that India is a large market for green energy, and projects tremendous opportunity to growth in green business (Fig. 4.3). Hoffman (2004) makes a compelling argument in this regard, suggesting that corporations remain agnostic about climate science and attach low significance to their 'social responsibility of protecting the global climate'.

Hilgartner and Bosk (1988) argue that political contexts impact the collective definition of problems, social or otherwise, and the political response needs to be viewed in that light. This points to the restrictive rural-agrarian overhang (Fig. 4.4) in political discourse. The making of political capital in the climate debate out of India's agrarian overhang, operating alongside the commercial impulses articulated via the *entrepreneurial self-identity* situates itself in the above discussion and in expert opinions (Fig. 4.5).

Putnam (1998) posits that a national government draws its relevance by, on the one hand, at the domestic level, seeking power by building effective coalitions of disparate interest groups, and on the other, at the international

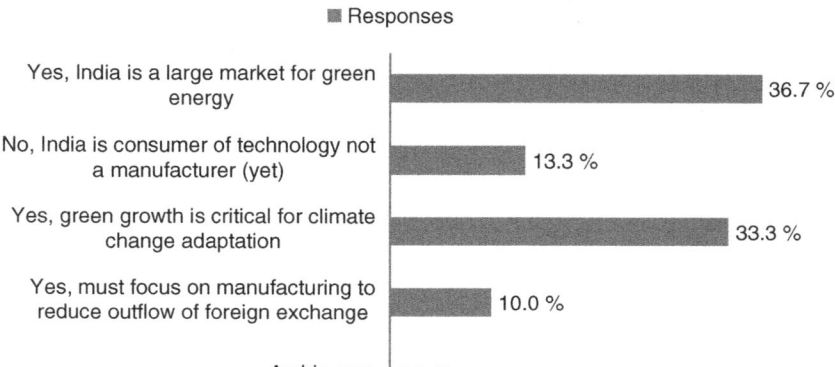

Fig. 4.3 India's entrepreneurial identity

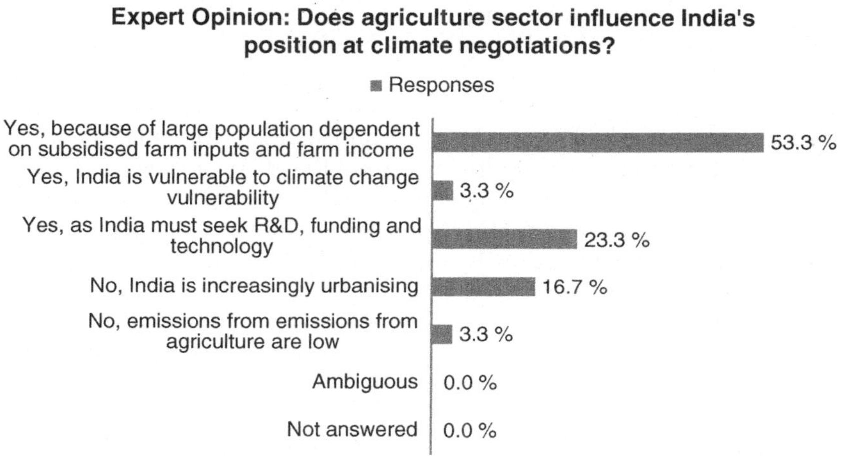

Expert Opinion: Does agriculture sector influence India's position at climate negotiations?
Responses
Yes, because of large population dependent on subsidised farm inputs and farm income
53.3 %
Yes, India is vulnerable to climate change vulnerability
3.3 %
Yes, as India must seek R&D, funding and technology
23.3 %
No, India is increasingly urbanising
16.7 %
No, emissions from emissions from agriculture are low
3.3 %
Ambiguous
0.0 %
Not answered
0.0 %

Fig. 4.4 Interview analysis: India's rural identity

level, by trying to maximise its capacity to cater to domestic pressures while reducing any adverse fallout of international developments. It is within this negotiated space that all international agreements fructify. Torn between divergent ideologies and contrasting political compulsions of various coalition partners, coalition governments of the past in India had very little

Expert Opinion: How will India manage sustainable, inclusive growth and generate employment?
Promote less capital intensive industries
13.3 %
Prioritise development of industries with a focus on energy security
30.0 %
Maintain right to develop and grow in International forum
6.7 %
Build manufacturing sector and the green economy
43.3 %
Others
3.3 %
Ambiguous
0.0 %
Not answered
3.3 %
Responses

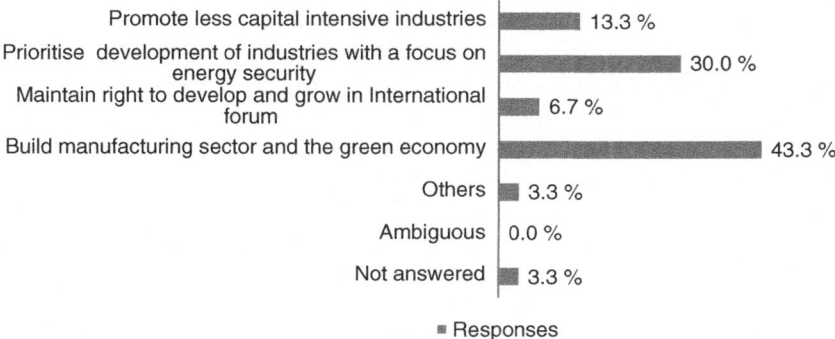

Fig. 4.5 Interview analysis: India's entrepreneurial identity

leverage space to negotiate and conclude any meaningful agreement in climate change. However, this landscape has changed.

The elasticity and brittleness of the public debate on climate, and the ease with which the government of the day is able to introduce powerful alternate narratives built around poverty and agrarian overhang, points to limited interest and understanding in the public sphere on climate change. In such contexts where voting publics have incomplete information, Wolinsky (1994, as cited in Sprinz and Weiß 2001), posits that governments concluding international environmental agreements are perceived to be more effective than ones that don't. Sprinz and Weiß (2001) align with this argument and suggest that in places where domestic audiences do not view climate change as a critical issue, national governments have significant leeway in concluding agreements with difficult terms.

Creating Leverage Within a Monolithic Identity

India is the sum of her states. Each of these states has its own cultural, political and social doctrines and dogmas, ironies and paradoxes, which together constitute the idea of India—a monolithic identity defined by its welcoming diversity. Outside of this framework, individual states have their own unique identities, operating out of a set of local impulses and imperatives, particularities and challenges. Yet, India's rural agrarian reality, and the rain-fed nature of the majority of its agriculture, remains a common unifying strand. Untaxed, and indulged with a host of unsustainable subsidies, India's farm sector enjoys high preference in the government's priority set.

However, the climate-dependent nature of India's agriculture sector makes her particularly vulnerable to the fallouts of climate change, especially in view of the challenging state of long-planned mitigating infrastructures such as interconnected rivers, extended canal networks, crop storage and logistics facilities, flood management mechanisms and drought control systems.

Adverse farm output and concomitant economic loss to the agrarian majority in turn impacts the political discourse, which makes governments very sensitive to the cause of the farm sector. The collective response of the people, responses by individual states and the response of the federal government towards climate change need to be viewed within this context. However, even as the federal government makes adequate representations subnationally, via subsidies, support prices and concessions, and supranationally, by invoking India's rural-agrarian overhang, transformative action for change has to come from the individual state.

CHAPTER 5

India's Official Stance

Abstract This chapter explores the position statements made by lead negotiators and political leaders and how they reflect the debate around the identities. It particularly focusses on key speeches made at pivotal moments of the international climate negotiations over the past 40 years and explores the evolution of these identities over time.

Keywords Conference of the Parties · Speeches · United Nations

The central hypothesis of this book is to explore if India struggles to tackle energy, development and climate change imperatives together due to the differing drivers it needs to accommodate when formulating responses to these challenges. The previous chapters help us develop the contours of the identity-based climate conversations in specific segments of the Indian media and among experts engaged with the subject. The other key enquiry of this research was to determine which identity or identities are seen to dominate India's position at the international climate negotiations.

It must be stated that while the development narrative may have been true for all of the Indian groups, geographies and stakeholders, these official pronouncements represent the perceived self-identity of the Indian collective in the view of just one stakeholder group, the Government of India (the federal national government). It would therefore be interesting to see if these are informed by the larger ecosystem of

© The Author(s) 2017
S. Saran, A. Jones, *India's Climate Change Identity*,
DOI 10.1007/978-3-319-46415-2_5

debate. Do the multiple Indian identity aggregations based on development realities as discussed previously find resonance with the official Indian position? And if they do, where does the balance of narrative reside? Poverty, opportunity, international leadership or ambiguity? Or is the result a cacophony of cross-currents emanating from competing needs and visions of 'itself' determined by one stakeholder, the government?

Here we review India's stance at climate change summits and uncover the core impulses of its position and also test whether there may be a degree of correspondence between these positions and Indian development landscape. The statements themselves are not unknown. Indian statements are in the public domain; there is nothing novel in their examination and many would have already subjected them to multiple levels of interrogation. What the research seeks to do here is to locate the Indian position within the unique development-based identity framework developed in this research and to then try and decode the meaning embedded within simple messages that may have eluded most, disappointed many and confused even more.

To conduct the textual analysis of the speeches made at the climate change negotiations, a carefully designed selection process was adopted. First, a set of speeches were selected based on the period of the research, that is, from 2009 to 2012. The other four speeches selected themselves as they were delivered at what were arguably the most significant meetings in the history of climate negotiations. The final sample set included India's official statements from the United Nations Conference on Human Development (UNCHD) 1972, the Earth Summit 1992, the World Summit on Sustainable Development 2002, the 13th Conference of Parties, Bali 2007, the 15th Conference of Parties, Copenhagen 2009, the 16th Conference of Parties, Cancun 2010, the 17th Conference of Parties, Durban 2011 and the 18th Conference of Parties and Doha 2012. Other speeches at the Conference of the Parties (COPs) where India was represented by a common speech from BASIC (Brazil, South Africa, India and China) or Group of 77 countries (G-77) groupings were left out.

With the help of the recurring keywords, ideas and phrases and taking cue from the suggestions (on Indian identity) culled from the previous chapters, an interrogative framework and coding schedule was prepared. This interrogation was then employed to analyse all of the speech texts.

As per this analysis, Fig. 5.1 shows the frequency of occurrence of some key themes in the eight official statements analysed. These include poverty eradication, agrarian dependence, food security, climate-induced

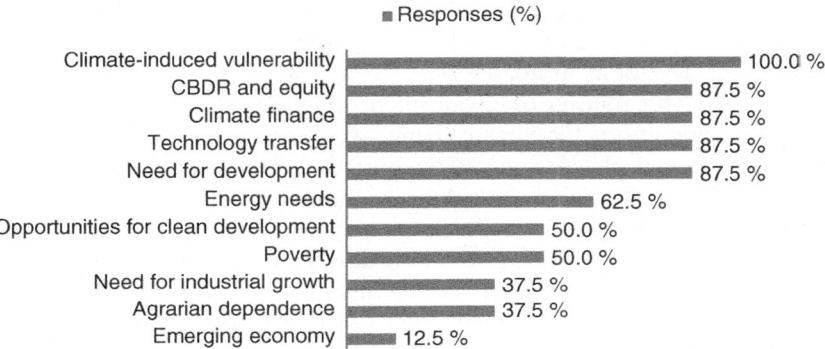

Fig. 5.1 Key themes from the Indian Official Speeches made at the Global Climate Change Forum

vulnerability, energy security, growth and development imperatives, transfer of environmentally sound technologies, institutional arrangements and funding for meeting the mitigation costs, principles of common but differentiated responsibilities (CBDR) and emerging economy.

Poverty reduction or eradication, as a major cause of concern for India and other developing countries, featured in 50% of the speeches analysed. At the UNCHD in 1972, Prime Minister Indira Gandhi strongly rooted this Indian stand noting: '... that environment cannot be improved in conditions of poverty. Nor can poverty be eradicated without the use of science and technology... The environmental problems in developing countries are not a side effect of excessive industrialisation but reflect the inadequacy of development' (Gandhi 2008).

Only 37.5% of the total speeches analysed mentioned food security or agrarian dependence as being important for India. Agriculture was mentioned within the broader ambit of vulnerabilities due to crop failure, food security and democratic governance systems. Adaptation was mentioned to be the foremost priority. All speeches referred to climate-induced vulnerabilities as a challenge for a developing country like India.

Of the total, 62.5% of the speeches mentioned India's large unmet energy needs for a vast population.

Out of the eight official statements analysed, 87.5% of the speeches emphasised giving the developing countries freedom to meet their growth and development objectives. It is important to note that the 12.5% of the speeches that did not mention these developmental aspects focussed more on the achievements made by India in portraying its commitment towards sustainable development goals.

Provisions for transfer of environment-friendly technologies for carrying mitigation efforts for climate change were mentioned by 87.5% of the speeches studied. These speeches exhorted the developed countries to acknowledge and support the voluntary commitments of the developing countries towards sustainable development by facilitating transfer of clean technologies. The arguments were fairly direct and clear, and therefore, no ambiguities were noted in these statements.

About 87.5% of the official statements made reference to the argument that cost of mitigation and adaptation in the developing countries should be borne by the developed countries. For instance, in 2012, at the 18th COP in Doha, the then Minister of Environment and Forests, Jayanthi Natarajan, stated: 'Simply creating the institutional arrangements like the Green Climate Fund does not help as its coffers are still empty. This brings us to the need for a clear roadmap on provision of finance for 2013–2015, 2015–2017 and then up to 2020. Public finance should be the primary source of fund for climate finance' (Natarajan 2011).

Out of the total, 87.5% of the speeches put emphasis on the CBDR and equity. During the World Summit on Sustainable Development, 2002, the then Minister of External Affairs, Yashwant Sinha, said: 'Sustainable development is a unifying philosophy. It was born of our combined idealism at Rio where we had pledged, each one of us, on the basis of our common but differentiated responsibilities and capabilities to act in a concerted manner for the greater good of mankind and our carrying planet' (Sinha 2002). The principle of CBDR and equity was observed to carry forward in most of the speeches made during the climate change negotiations by India.

Only 37.5% of the speech texts implicitly mentioned need for industrial growth, job creation or power generation; 50% of the speeches discussed India's entrepreneurial identity by way of mentioning India's commitment towards meeting energy efficiency goals and deploying alternative sources of energy.

Being a more recent development, only 12.5% of the speeches mentioned India as an emerging economy. During the 16th COP in Cancun 2010, Jairam Ramesh, the then Minister of Environment and Forests, stated India to be one of the fastest growing economies and declared its commitment to be among the most responsible in ensuring a Green Domestic

Product-based economy (Ramesh 2010b). In other speech texts, however, there wasn't any mention of India's emerging nation identity.

Evolving Identities; Congruent Identities

This textual analysis offers two significant insights: First, it helps us to observe the evolution of the Indian positions (official) on the climate debate over time, an indicator of evolving self-identification, and thereafter, it also helps test the congruence between the aggregate identities gleaned from the analysis of India's development – energy landscape and from the distribution of focus in the official statements/speeches made at the global climate change forums. Table 5.1 shows the coincidence of these themes within these official texts with the six identities.

From the review of the data, policy and academic literature dealing with India's development landscape, and juxtaposed alongside the key self-identities flowing from Indian official statements at the international arena, these six key identities resonate closely with the Indian macro-propositions at these global negotiations and meetings.

In an effort to bring together the industrialised and developing nations together to outline the 'rights' of humanity to a productive environment, the UN Conference on the Human Environment 1972 witnessed India's active participation. Priority of poverty eradication and development growth for India was put forth by the then Prime Minister Indira Gandhi (Gandhi 2008). Not only did the Stockholm conference put environment on the political agenda but it also placed the poverty eradication agenda at the heart of the climate debate for times to come. Undeniably, India had a major role to play in that. India asserted that 'development' is one of the primary means of improving the environment for living, providing food, sanitation and shelter. It questioned the real cause of environmental destruction and argued that pollution is not just a technical problem and no amount of environmental conservation efforts will ever be successful unless poverty is effectively dealt with. Highlighting the income disparities and economic asymmetries between the participating countries at the conference, it can be assessed that India portrayed the *rural* and *developing nation identity* strongly in 1972. It laid the foundation of India's consequent position in Kyoto (1992) re-emphasising attention towards poverty eradication and conforming to principles of CBDR, environmental justice and equity.

Table 5.1 Textual analysis findings

Identities	Interrogative framework	Agree (%)	Disagree (%)	Not mentioned (%)
Rural	Poverty eradication is a major concern for developing countries (India)	50.0	0.0	50.0
	Agrarian dependence (subsidies, labour, etc.), food security is a major concern for developing countries (India)	37.5	0.0	62.5
	Climate-induced vulnerability is a major concern for developing countries (India)	100.0	0.0	0.0
Energy Security	Energy security is a major concern for developing countries (India)	62.5	0.0	37.5
Developing	Need for development and growth in developing countries must be respected	87.5	0.0	12.5
	Transfer of environmentally-sound technologies by developed countries should be facilitated	87.5	0.0	12.5
	Cost of climate action should be borne by rich countries	87.5	0.0	12.5
	Principles of common but differentiated responsibilities and equity must be pursued in climate change negotiations	87.5	0.0	12.5
Industrial	Need for industrial growth, job creation and/or power generation	37.5	0.0	62.5
Entrepreneurial	Opportunities and potential for clean development, renewables and energy efficiency	50.0	0.0	50.0
Emerging	India is an emerging economy	12.5	0.0	87.5

At Kyoto in 1992, besides presenting India as a country with a low level of industrial development, P. V. Narasimha Rao, the then Prime Minister, pledged India's capacities to the collective action to mitigate climate change (United Nations 1993). Its capacities were limited to traditional knowledge systems that exist in India in the form of herbal medicines, water harvesting and management and some other conventional technologies. A clear balance was maintained between India's vulnerability and competence in tackling the climate change issue. More distinctly, the energy security concerns and development of alternate forms of energy were robustly put forward for the first time. However, renewable energy development, eco-friendly technologies, use of biotechnology for conservation and health improvement, and traditional knowledge were the key focus of the speech made in 1992. It would be safe to assume that 'technological research and development and transfer' had the spotlight indicating the *developing nation identity* of India.

As evident through the textual analysis, India re-stressed upon tackling the poverty issue and sustainable development at the World Summit on Sustainable Development, 2002. India put into focus the concern towards unsustainable consumption, governance, debt reduction and subsidies. It could be felt that India was projecting the vulnerabilities both due to a large agrarian-dependent population and challenges faced in pursuit of growth and development targets. Within the broader ambit of sustainable development, India re-emphasised its *developing nation identity* and demanded financial assistance, CBDR principles and eco-friendly technologies.

By 2007, India seemed to be more focussed on specific issues highlighted broadly at the 1992 and 2002 conferences such as technologies for mitigation and institutional arrangements for funding. India practically set the tone for the roundtable discussion on technology transfer. Steering the debate towards research and development of mitigation technologies, their adoption and implementation, India pressed for a country-driven approach. It stood out as one of the stronger voices, representing the case for the South demanding global action on technological development. Arguably, the speech outlined what India or the other developing nations sought from the global community in mitigating the impacts of climate change. However, it is difficult to assign a particular identity India portrayed at the Bali COP 2007. The dominant identity for 2007 lies in between an *emerging* and *developing nation identity*.

The Copenhagen Summit, 2009, aimed to reach an agreement on binding commitments after the end of the Kyoto Protocol commitment period in 2012, but it failed (McKibben 2012). India professed the principles of the UNFCCC (United Nations Framework Convention on Climate Change), in particular the principle of equity and CBDR and respective capabilities. In depicting the responsibility undertaken towards a global action, India confidently presented the National Action Plan on Climate Change and the set target of installing 20,000 MW of solar power by 2022, along with an addition of six million hectares of forests in the coming few years. While the need for respecting the space and needs of developing economies were acknowledged in the speech, agreement on voluntary targets for reducing emissions were submitted. This statement was considered to be constructive by some who averred that India has made a significant proposal that would help in bridging the differences within global action required on climate change (Jaiswal et al. 2010; Hallding et al. 2011). However, due to its stand in demanding development space, India's developing nation identity was prominent again in the 2009 negotiations.

In 2010, at the Cancun COP, the official speech introduced India's Green Revolution story putting agriculture, technological development and innovation at the forefront. It was followed by an update on India's efforts domestically, including its emission reduction targets, diversification of energy fuel mix and aggressive strategies on forestry and coastal management and regional cooperation on climate change. These actions were presented to the forum perhaps implying India's potential of playing a leadership role at the negotiations. For the first time, India acknowledged its emerging economy status. 'Environmental stewardship demands responsive leadership,' said Jairam Ramesh, leader of the India Delegation at Cancun. Alongside this, he facilitated the debut of India's *emerging nation identity* (Ramesh 2010b).

Notwithstanding, in 2011, Minister Jayanthi Natarajan in her speech brought back India's *developing nation identity* and put the emphasis back on poverty reduction and eradication, vulnerabilities and large unmet energy needs (Natarajan 2011). Much of the focus was placed on cost to combat climate change and the importance of technology transfer. She emphasised on the need to acknowledge the effective mechanisms to deal with 'loss and damage'. The pendulum had swung back and the poverty narrative made a strong return.

At Rio+20 Conference in 2012, Mira Mehrishi, Special Secretary, Ministry of Environment, Forests and Climate Change, proposed India's ambitious targets for renewable energy, energy efficiency and lower carbon growth pathways, insisting on financing and greater international support (Mehrishi 2012). Her articulation reflected India's *rural, energy* and *developing nation identity*. The Rio+20 Conference on sustainable development saw India as one of the dominant voices explicitly asserting that poverty eradication remains the greatest global challenge and India's foremost priority.

Discovering the 'I'

The formation of perceptions on climate change is a complex organic process. Here we have tried to identify the most prominent drivers of this perception formation. The methodology for doing so has involved an assessment of media discourse, interviews of key stakeholder groups and interrogations of speeches and policy texts emanating from India's climate change positions in international forums. The result of this exercise has been that a number of seemingly contrasting and sometimes contradictory perceptions can be found both between and within each of these categories.

To the casual observer, this may suggest the lack of a pattern or indeed a cohesive Indian position and vision. However, on critical analysis, this suggests that the interpretation of results, through a cognitive prism, must assume that the whole is greater than the sum of its parts and that 'whole' is what this study seeks to describe.

We see a confident India taking on a more progressive leadership role in global forums following from its *emerging nation identity*. As a 2 trillion-dollar economy, India has begun to outgrow its G-77 shoes, even as it has a range of development concerns left to address. The speech analysis in this chapter has shown that the *emerging country identity* only manifested after the COP in Copenhagen. In the years preceding, *rural* and *developing country identity* and the *industrialisation* imperatives were highlighted repeatedly. And in the following years, the *entrepreneurial* and *energy security identities* assumed prominence. These patterns show that the *emerging nation identity* is not one that posits that India has overcome its development goals. It is one that simply affords India the chance to manage its poverty by focusing on resources management and entrepreneurial opportunities and acquire greater clout on matters of global governance in order to safeguard its major development concerns.

We also find the usefulness of the green economy being portrayed. Observed in the articulation of the potential and opportunities offered by the renewable energy sector are efficiency improvements that can assist the agrarian and industrial imperatives in the form of entrepreneurial opportunities that the new low-carbon economy allows all comers. Therefore, economic rationale is central to this narrative. While the political class may see the green economy as a means to cater to populist impulses of employment generation or for furthering their global standing, the media and professional stakeholders may see it as a means for unleashing of business opportunities from the grassroots to the stock exchanges, reflecting the global mood. What is quite apparent is that this 'green base' is now a constant and has emerged from the business case offered by these new sectors and the economic rationale of change. Each stakeholder and indeed the narrative on the whole have to accommodate this 'green reality' even as they posit their own perceptions on other realities. This green and low-carbon narrative, therefore, forms the base of an imagined triangle, deepening and strengthening as economics overwhelms politics.

While the agrarian overhang is witnessed consistently, the reality of India's poverty, and the deployment of this reality by officials at international forums, is not as strong as some would expect. In this chapter itself, for example, the analysis shows that food security and agrarian dependence are not answered as prominently when Indian leaders interact with the world outside.

In fact, it would seem that in order to transit from this *developing country identity* to an *emerging country identity* and finally on to the global high table itself, Indian negotiators and official interlocutors are underplaying (relatively) the continuing saga of poverty that defines much of India's development narrative. And consequently, it shapes what they seek from these negotiations. This is not to say that poverty is not invoked for the purposes of negotiations, it clearly is, however the use has a different texture to what a similar proposition would be, had it been voiced by a poor African or island nation.

The contemporary Indian invocation of the income-poverty narrative is on a broader question of equity and differentiation between developed and developing countries and for the primary purpose of discovering legroom for its industrial and entrepreneurial personas. Therefore, despite sharing relatively similar human development indices (HDI) with sub-Saharan Africa, speech analysis points towards the fact that India is much less likely to raise issues of its own HDI in global forums than African countries for which HDI often forms the pivot of negotiating positions (Mburia 2015).

In other words, rather than hiding behind poverty, India, it seems, struggles to hide its poverty at global forums.

PART III

The Modi Factor and India's Future Identity

CHAPTER 6

Looking Ahead

Abstract The final chapter reflects on the identities and the Paris Conference of the Parties in December 2015. It includes a discussion on the possible implications of the identities presented in this book and reflects on what role India may play in future agreements and the implementation of those agreements. In particular, it includes new analysis and reflections on agreements made in Paris and the role of Prime Minister Modi.

Keywords Prime Minister Modi · Climate justice · Paris agreement

The role of media in the emergence of social movements, political discourses and civic engagement in governance is undeniable. As a result, the form, type and structure of mass media have been studied in detail in post-industrial societies and more recently in developing and emerging economies. Media has been blamed for causing civic 'malaise', that is news practices that hinder civic engagements and political activism (Norris 2000). On the other hand, media is seen as mistakenly construed for being the exclusive reason for such malaise, overlooking underlying political and social factors at play (Norris 2000). Therefore, it is safe to assume that media functions through complex processes in exerting diverse impacts on different political and social systems. In this connection, it can be posited that reporting and interpretation of news forms a strong basis to examine the influence of news media.

However, many of us confuse the noun for the verb and the act with the actor. When we discuss media coverage, are we discussing an institutional output or an individual's expression? Do editorials shape coverage or do individual biases and self-identification prove to be more decisive. For someone who has studied media over some duration, one of the critical factors is individual journalistic biases in presenting the news. Socio-cultural context, political orientation, institutional influences and other structural factors are said to influence journalistic biases (Mazzoleni 2003). Individual biases are usually introduced in the media through a set of language and tone of the reporting (Lakoff 1996), and these are seldom an outcome of editorials.

The US Media Research Centre has conducted extensive surveys to understand perceptions around credibility of newspapers and journalism and it has found that a majority of the American audience find the media biased (in one way or the other) and doubt the integrity of the reporters. The Pew Research Center collected similar responses through poll results in 2008 regarding perceptions about the authenticity of reporters and asserted that there has been little change in such perceptions over a decade. Unfortunately, comparable studies are missing in India even though some attention is being paid to politicisation of media. For instance, the 2010 controversy of Nira Radia tapes brought journalistic biases into focus. Controversy encircled acclaimed journalists who were blamed for manipulating the news and disrespecting media ethics.

While certain trends in reporting are visible in newspapers as assessed in this study, it is hard to put this down to any editorial bias. Indeed, what this may be is just how one reporter or sets of reporters identify with the issue and deliberately or unwittingly end up shaping the discourse significantly. These can be confirmed by some counter-intuitive findings, which see more left of centre newspapers making a business case aligned with market economics and more right of centre papers making a case for socialist approaches. This is one element that the study has not really ploughed into and would make a fascinating research for the future.

Of course the future role of media as an actor is changing rapidly. On the one hand, the scope and potential of the online mainstream news is being challenged by increasing market competition, constant demand for dynamic and fast-moving content, and interactive platforms. On the other, changing social realities such as dwindling culture and habit of conventional reading, despite increasing literacy rate, has been challenging to both print and formal online news media. With the advent of Twitter and other social media platforms, the world is discussed, debated and judged in 140 characters.

As digital media technologies evolve, they will no doubt impact every aspect of information gathering, reporting and sharing. Definitional boundaries of online journalism and new media as offered above are fluid and juxtaposed. Mainstream media is fast adapting to new media. This has involved a fundamental reassessment of interaction between new media (including digital journalism) and institutions (social, political or economic) through which commentaries and actions unfold.

However, the second issue around the growth of this medium is far more important and will become an enduring reality of this century. Most now agree that the extent of popularity and speed of penetration of online social networks have been exceptional. It is expected that by 2017, global social network users will increase to at least 2.55 billion, constituting nearly 40% of the global population and with an associated impact on many more. Increasingly, the vastness and efficiency in interaction of these social networks will not only shape the discourse and social structures, political participation and politics, activism and agitation but also change the conception of collective identity (Hanrath and Leggewie 2013). In the future, the whole idea of lived experience and community may be based on virtual histories, digital citizenship and cyber societies, who are already aggregating in manners different to the land and sea boundaries of nations and sometimes distanced from the social–economic realities of their locality. Therefore, the whole notion of 'national identities' could become less relevant.

New Delhi Is Not India

The Union of India is comprised of 29 states and seven union territories (UTs). These states, by virtue of the Indian Constitution, are empowered to decide on a number of critical policy areas. India is therefore a federal democracy. Decision-making on subjects of local, regional and global importance, are all part of the federal matrix. The states have legislative powers over critical subjects such as land, water, energy and agriculture. Since these subjects in turn will undeniably be impacted by climate change and their management will in turn shape climate agendas, the 'monolithic identity' of India that obscures the role of the states is problematic and requires future research.

The Indian Constitution provides each state with the freedom to execute laws on subjects classified under the 'State List' (List II). Article 246 of the Indian Constitution lists 66 subjects, over which states have the power to draft laws. Similarly, the Union List has 97 subjects over which the Parliament has exclusive legislative powers. These include defence, foreign affairs, interstate transportation and interstate rivers. The concurrent list is comprised of 52 subjects (forests, protection of wildlife, mines and minerals, etc.) with shared jurisdiction between the States and the Union Government. However, Article 253 of the constitution has empowered the Parliament to make any law that is deemed necessary to implement international agreement, treaty or convention, foregoing provisions for state subjects.

In the early 1990s, the emergence of coalition of parties in the central government and economic liberalisation policies triggered a new impulse of decentralisation of power to the states. The growing significance of regional political parties is a testament to this trend. In 1992, Panchayati Raj Institutions and Urban Local Bodies (colloquially called the 'third tier of governance') were empowered by Constitutional amendments (73rd and 74th), leading to further devolution of powers and jurisdictions (Singh 1994).

Economic power and capacity of the states can be evaluated from the trends in plan outlay and expenditure in the Five-Year Plans of India. From the first Five-Year Plan in 1951–1956 to the recently completed Eleventh Five-Year Plan (2007–12), planned outlay for development programmes, projects, schemes on capital or revenue account have been much higher for centre than states/UTs. However, expenditure shows an important trend. Expenditure of the states/UTs, as compared to their sanctioned outlay, has been much higher throughout almost all the Five-Year Plans except for the second, sixth, ninth and tenth Five-Year Plan periods.

The Finance Commission of India, established under Article 280 of the Constitution to assist in financial relations between Centre and State, recommended increasing the share of states in the central taxes. This is indeed an important step in the devolution of funds and greater autonomy of the states. The terms of reference for the 14th Finance Commission Report also included sustainable development, inclusive growth, ecological considerations and environment and climate change sensitivities. These are therefore important determinants of financial flows from the centre to the states. For instance, the Finance Commission incorporated devolution of central taxes from forest cover – assigning 7.5% weight for determination of fiscal devolution to the states. Indian states with increasing political and economic empowerment and control over critical natural resources will become central and vital to climate action at the local and global levels.

Flowing from this federal reality of India, it would also be important to reemphasise the role of the vernacular media and its influence on setting the developmental and political agenda in India. The vernacular press has significant agency in aggregating concerns of local communities. Vernacular newspapers in India began as early as the 1760s, including *Bengal Gazette, Digdarshan, Aligurh Institute Gazette* and so on, depicting local and regional discourses (Akhtar et al. 2010). By the middle of the nineteenth century, newspapers were published in most of the popular Indian languages. However, in 1878, the Vernacular Press Act was passed to control the content and circulation of these newspapers by the British Government. The Act was repealed in 1881, and nationalist movements in the nineteenth century further pushed the development of local print media.

Today, the rapid growth trend of the vernacular media should be placed alongside the falling share of national English dailies. Increasing literacy rates, improved access to rural markets and changing macro-economic environment have led to substantial growth of vernacular print media in India, in the order of 10% annual average growth rate (FICCI-KPMG 2013). The launch of local editions and the gradual improvement in advertisement rates in rural markets have further increased the share of vernacular dailies in the total print media revenues. Large media houses such as *Hindustan Times, Dainik Bhaskar* and *Deccan Chronicle* have also started targeting location-specific reach. As a result, Hindi and vernacular dailies constitute more than 60% in the market share of print media (Kumar and Sarma 2015). The trend is expected to continue at 9% annual growth rate on account of rising consumption demand at the state level (Kumar and Sarma 2015).

It is evident that the significance and role of vernacular media will be critical in setting political and social agendas. Increasingly the vernacular media and the federal political culture will shape international engagements as well. Today there is a tendency in the vernacular press of not actively engaging with some of these global issues such as climate change.

This could and may change in the years ahead. However, an important caveat to mention here is that in the interviews conducted in this study, there is an agrarian overhang despite the fact that all people interviewed live in urban areas. This seemingly contradictory outcome can be attributed to the perceived importance of rural and agricultural development in India (among other factors) given the dependence of 110 million people on agriculture for generating livelihoods and positioned at the centre of vernacular reportage. Could the vernacular press also be influencing the agenda for the English language newspapers? How the rural and agrarian concerns are observed, transmitted/collected and reproduced by the English media would have a bearing on shaping the lived experience of urban dwellers.

The Modi Factor

Perhaps, an important factor in predicting India's climate action will be the role of political leadership. One way of understanding the influence of 'leadership' could be through analysis of the processes by which leaders are chosen within different social, political and cultural ecosystems. In the context of this study, it is important to acknowledge that political leadership is considered as a 'thick, potentially all-inclusive, subtype of social leadership' (Masciulli et al. 2009) and is a part of social process that fosters collective action (King 2002).

There have been various studies on the characteristics of the leader but they are mostly culturally biased (Helms 2014). Generally, effective leaders are thought of as charismatic, innovative, adaptive and simultaneously disruptive. Recent political leadership in India and the United States demonstrate the power of leadership in bringing increased attention to climate change. For instance, the current Indian Prime Minister, Narendra Modi, has made major efforts in articulating India's local climate action and therefore has set the benchmark for enhancing the country's propositional role in climate change negotiations. Within the first year in government, he announced deployment of 100 GW of solar energy by 2022 alongside expansive energy efficiency promotion programmes and projects for improving air quality and urban waste management.

In a recent speech at the UNESCO on 10 April 2015, Mr Modi said, 'Climate change is a pressing global challenge. And, it calls for a collective human action; and, a comprehensive response' (Modi 2015b). He emphasised the significance of harmonious institutional relationships, the role of Indian traditional knowledge and innovative technological fixes that can help the world deal with climate change. In New Delhi, while addressing the conference of State Environment Ministers on 6 April 2015, he further asserted India's leadership potential by stating that 'India should guide the world on fighting climate change' (Deccan Chronicle 2015).

President Barack Obama's leadership too is noteworthy. His particular leadership expertise is described as 'A leader for the "We" generation' (Jones 2015). Mr Obama's policies and initiatives on health care, jobs, religion and climate change have been much criticised and also appreciated for raising the pitch and attention of debates on these issues. It is suggested that strong leaders are motivated by encouragement, altruism and moral consciousness (Colomer 1995; Frohlich et al. 1971). His 'bottom-up, empowering' approaches to institutional processes in the United States (Colomer 1995; Frohlich et al. 1971), affecting sectors like health, businesses and education, have certainly had political impact.

To address climate change, Obama announced plans to cut emissions from power plants, building climate resilience and leading global efforts on climate change. From the very beginning of his tenure, Obama acknowledged the sense of urgency towards environment protection. In his speech at St. Paul, Minnesota, in 2008, he outlined his 'ambitions' and said 'we will be able to look back and tell our children that this was the moment when we began to provide care for the sick and good jobs to the jobless. This was the moment when the rise of the oceans began to slow and our planet began to heal. This was the moment when we ended a war, and secured our nation, and restored our image as the last, best hope on Earth' (The New York Times 2008). In 2014, Obama proposed the first ever 'Clean Power Plan' for setting and regulating pollution standards for power plants and announced joint commitment to reduce emissions with the second most significant energy consumer and emitter country, China (The White House 2015).

Given the current status of climate negotiations, the critical role of leaders such as Obama and Modi in defining external engagements of their nations – the 'leadership' factor – needs more detailed research. Leaders lead by uncovering new pathways and scripting new narratives. Sometimes they invoke moral propositions to effect practical outcomes. Perhaps when Prime Minsiter (PM) Modi re-invoked and popularised the term climate justice in the months preceding Paris, he was doing all of the above.

The phrase climate justice has been around for quite some time. Indeed, the first popular invocation of the term can be traced back to the Climate Justice Summit held in Hague in 2000, at the same time as the sixth Conference of Parties (COP 6). The Summit called for a 'climate justice'-based approach to solving the problem of climate change – an approach that would promote human rights, equity, labour rights and environmental justice, both across and within nations (Whitehead 2014).

The term climate justice has originally been an attempt to frame the challenges of global warming and climate change as one of rights – an ethical and political problem – rather than one that is solely environmental or physical. The impact of climate change on fundamental rights, on equality and on question of social justice all makes this framing a powerful one. Historical responsibilities, collective rights and differentiation in capacities are all a part of the climate justice framing.

The climate justice movement received another fillip when the Bali Principles of Climate Justice were adopted in 2002. The principles call for ecological debt, the responsibility of global corporations and the rights of indigenous people to be incorporated in any global response to climate change (Bali 2002).

Principally speaking, the term climate justice refers to two components – distributive and corrective justice (Savaresi 2016). Distributive justice has to do with distributing the burden of mitigation as well as capacities such as finance and technology. Corrective justice has to do with the question of loss and damage and the unequal impacts of climate change.

Following the publication of the Bali Principles, there have been other conferences and movements that have emphasised the importance of climate justice. The term, however, receded from the forefront of the climate discourse and had little relevance in official climate policy, confined instead to civil society groups, activists and environmentalists. This changed, however, when PM Narendra Modi spoke of climate justice in September 2015 at the United Nations Headquarters in New York, months before the Paris negotiations (Srivas 2015).

Consequently, the term gained fresh impetus and several Indian government officials began repeating the phrase, signalling their intent to push through the agenda of climate justice into the Paris Agreement. Indian Environment Minister Prakash Javadekar in November 2015 called for developed countries to free up carbon space and said, 'we want climate justice for the billions of poor of this world' (Firstpost 2015).

In the end, climate justice has figured in the preamble of the Paris Agreement, which notes '...the importance for some of the concept of climate justice' (UNFCCC 2015), when taking action to address climate change. The term is therefore now embedded in the future of the new climate regime and can be understood as a placeholder in the current agreement. Questions over climate justice and human rights are only likely to intensify in the years to come as the impacts of climate change deteriorate further.

PM Modi's intervention on climate justice in the lead up to Paris was a seminal moment. The impact of the reframing of the discourse around equity and differentiation considerations through the connotations around climate justice cannot be underestimated. Put simply, Modi achieved three things through the reintroduction of climate justice in mainstream climate debates:

1. The gridlock around common but differentiated responsibilities (CBDR) and respective capabilities (RC) was removed. The sharp Manichean-like designing of equity that was based solely on historical responsibility was never likely to be politically viable in the Paris Agreement. At the same time, developing countries were always unlikely to relent on their arguments around differentiation. By invoking the concept of climate justice, Modi allowed climate debates to move past the disputes and divisive politics around CBDR, while retaining its ethos, principles of differentiation and moral grounding.
2. The concept of climate justice enabled climate debates to move from a north/south bipolar tussle to a global, inclusive agenda. This is because climate justice refers to all people everywhere, including citizens of developed countries who are victims of inequality and social injustices and cannot be expected to bear high levels of responsibility, while rich citizens in developing countries are excluded from taking action. The spectrum of debate was therefore widened and climate justice enabled an inclusiveness, hitherto absent in discussions around equity and differentiation.
3. Last but not least, Modi generated a narrative that the Paris Agreement and all its parties could rally around. A propositional rather than oppositional India emerged during the negotiations, much to the relief of everyone who had long grown tired of India's trade union-style politics at global forums. A forward-looking,

agenda-setting India looking to build consensus had evolved, in sharp contrast to the reactive Indian position at previous COPs.

To operationalise the principles of climate justice, India can contribute in two ways. First, India can push forward the argument that development, adaptation and mitigation should be seen as same sides of one coin. For far too long, development has been associated with the 'right to pollute', a framing that has in a sense polluted the climate discourse. We need to move beyond the idea of a right to carbon space for some and instead talk of the right to life and sustainable development for all. That itself creates a new positivist, forward-looking paradigm that all countries will be happy to associate with. In fact, the placeholders of climate justice and human rights in the Paris Agreement offer us an opportunity to make this shift in narratives. This reframing is also important because adaptation and development will lead directly to mitigation. As India alleviates poverty and increases the purchasing power of its people, clean energy options will witness increasing uptake, contributing to mitigation action. As has been highlighted previously, India will grow coal to go green.

The second aspect that must be considered in any operationalisation of climate justice is the question of how do we apportion our carbon space equitably within countries. Demanding a share of the carbon space from developed countries who have occupied more than their fair share will be counterproductive if that only leads to elites in developing countries such as India doing the same within nations. To avoid this, India can take a few important first steps:

- Follow CBDR within India, requiring rich states and cities to develop innovative mitigation methods, including through 'Green Building' Initiatives, improvement in public transport infrastructure and adoption of energy efficiency schemes by the affluent, each of which is already at various stages of implementation at the central and state levels.
- Initiating a universal agreement on corporate emissions mitigation that would involve large Indian companies on equal footing with developed country corporations and mandating sectoral efficiency goals for these large corporations.
- A decadal review of India's development status within global climate forums, as no exception should outlive its rationale.

Some of these action areas are already being pursued by the government. India's rich should not be allowed to hide behind its poor, and similarly, developed countries must not be allowed to hide behind India's rich. This is critical to the successful operationalisation of climate justice principles in our climate action going forward.

India's reintroduction of climate justice into mainstream climate politics has served to ideologically evolve the debates around equity and break the deadlocks that were becoming associated with the principles of CBDR and RC. The north–south divisions have been cleaved to unify behind a focus on human values and common ethical questions.

This is not to say that this reframing will be a panacea for all things climate. The challenges of operationalising climate justice will be as profound as the challenges of operationalising CBDR. Distributive justice in the context of technology transfer and flows of finance will continue to be debated. Corrective justice in the context of loss and damage and compensation is only likely to be more controversial in the years to come as entire countries face the threat of going underwater due to actions undertaken by others.

However, we may have moved beyond the divisive politics of CBDR and reached a global deal on climate change with all the big four emitters on board whilst retaining the moral principles and ethos that created CBDR and RC, through the concept of climate justice. India's own approach to climate justice must be unquestionable as it argues for global outcomes linked to the same. The new narrative that has been generated about India's efforts on climate justice must be supported through the momentum of increasingly ambitious and inclusive action. Its domestic policies must demonstrate that climate justice protects lifeline not lifestyle needs and that the most vulnerable and poor are to be safeguarded from the impacts of climate change. When the consistency of this approach gains ground across nations, equitable and ambitious outcomes cannot fail to be far behind.

Apart from the reference to climate justice, PM Modi has also looked to create a new narrative around India's climate action. India's ambitious goal of increasing wind and solar energy capacity in the country to 175 GW by 2022 as set by Modi has become the new symbol of India's efforts on climate change, a far cry from the days when the focus was on India's growing emissions and intransigence over its use of coal. While the reality of coal usage continues to remain the same (and is much less than that of the average Chinese or American on a per capita basis), the narrative and imagery around India has visibly changed. Even though India's nationally determined contribution makes it clear that its energy consumption is likely to grow four times, India's per capita emissions are unlikely to ever cross that of developed countries. The bottom line is that India will grow coal to go green. Yet, now it is the second part of that equation that is receiving attention.

This recasting of old themes that has been undertaken is a clear directional and messaging effort by PM Modi, and is in line with the requirements of leadership and vision required to facilitate a fair appreciation of India's role in the global effort on climate change. Shifting the locus of debates away from coal and twentieth-century development prerogatives (while continuing to pursue them unabashedly) to technology transfer, innovation and clean energy growth have led to the redefining of India's role in global climate politics.

Overall, given all the constraints that India faces in tailoring an international agreement that is aligned with its own self-diagnosis, the ability of political leadership to capture the media space in many ways offers hope in the contemporary context. It represents a clear opportunity for PM Modi to propose a win–win deal to the parties to the convention, one that offers India and other similarly placed developing countries with the opportunity to ensure both lifeline and lifestyle growth while simultaneously ensuring global action. It represents an opportunity to align India's international positioning with its growing aspirational class while simultaneously protecting the interests of those at the bottom of the socio-economic pyramid. It can also be deduced that political leadership by India may give a clearer shape to its identities.

In Sum

In the global climate discourse, India's stance has been defined by the interplay of multiple identities and subidentities that simultaneously jostle for space. India's official commentary at international climate change discussions from the origins of its engagement on this issue in 1972 to contemporary times has been built around a minimalistic narrative. It is built around rural–urban economic inequity, agrarian imperatives, lack of access to basic resources such as energy, even as India committed itself to responsible climate conduct. These trends are clearly observed in the policy statements and speeches of Indian leaders.

However, the media narrative and expert opinions have been more expansive in articulating priorities. Even as issues such as those that have been consistently visible in the speeches of Indian leaders have been reaffirmed by expert opinions or media reportage, their set of concerns also encompass elements such as the growth of the renewable energy sector and the green economy in general, and India's growing role in the international sphere and the commensurate responsibilities.

On analysis it becomes evident that the media space is brittle and available for capture. This is clear, for instance, in the shifting sands of media discourse in the run up to, during and after the COP summits that are covered in this research. For instance, both in 2009 and in 2012, the *developing country identity* was amplified during the COP time period, compared with the pre-COP period. It can be posited that such a change was a result of infusion of the political direction within the media reportage, in turn giving India's leaders greater room to manoeuvre for their negotiating stance.

In this regard, the trends in the media reportage pre, during and post the COPs in 2009 and 2012 need some examination. The pre-COP discourse, even while acknowledging the income and opportunity disparities within India, situates itself largely around an aspirational narrative. Aspirational identities such as *entrepreneurial* and *industrial* find strong resonance in the pre-COP phase. However, as the conferences get underway, a more minimalist agenda articulated through *developing* and *rural identities*, the dominating strand of India's official representation at the climate debate, finds sharper resonance. In India, where the media is highly trusted to present an objective worldview on technical subjects such as climate change, almost at par with climate scientists themselves (Leiserowitz and Thaker 2012), such a drawdown has its own ramifications and can perpetuate systemic policy inertia that can only be disrupted by strong leadership.

For instance, in the absence of such disruptive intervention, the Indian Minister positioned the Indian approach by relying on its *rural identity* at COP-19 in Warsaw. She ensured continuity of the minimalist agenda by stating 'poverty eradication stands as our foremost priority. We have huge social and developmental constraints and have to address large unmet energy needs of our vast population....' (Natarajan 2013).

Perhaps the first signs of India rethinking its climate position were visible during the COP-20 in Lima (Jones 2015). Some argue that the firewall between Annex 1 and non-Annex 1 countries was breached. The Indian Minister in fact boldly placed sustainable development alongside poverty eradication (Javadekar 2014).

What is also visible is that stakeholder groups often contradict themselves. For instance, young professionals interviewed as part of this research are optimistic about the green economy and the entrepreneurial ecosystem but not the renewable sector in particular. Their main locus seems to be jobs, yet not in the manufacturing sector. And at the same time they don't seem to want more openness in the economy. They are also *developing country* champions. Therefore, there is little by way of a cognitive consistency visible in the responses within the group. This closely mirrors the responses from across the stakeholder groups and gender/experience categories, as well as media reportage. Seen together they represent myriad perspectives and varied interests. They come together to comprise a whole that at times seems purposeful and at other times seems to be lacking direction.

In projecting multiple self-identities, India is neither being duplicitous nor hiding behind its poverty as some commentators seem to suggest. Instead, it is reflective of the dominant national sentiment that filters through as an aggregation of stakeholder perspectives. Some of this heterogeneity that turns to homogeneity at the aggregate level can also be explained through the role of the media in perception formation. It is clear that external factors affect perception formation since there is a wide array of opinions within stakeholder groups, such as young professionals, who would otherwise be expected to have overlapping loci of interests (such as jobs).

It is clear from the aggregate trends that India needs to strike a balance between climate and development. It needs to preserve its CBDR anchor while coordinating closely with countries that comprise BASIC (Brazil, South Africa, India and China) and in particular its giant neighbour China. It needs to take a leadership role in international discussions while continuing to work with the Global South. It needs to build coal sector efficiency even as it begins to bolster focus on renewable energy. These multiple identities may appear distinctive or even antithetical. Yet they are neither. Rather, like the multitudinous brooks and streams that meet to form a river where their identities merge, India's identities ultimately unite to one single purpose – to lift her teeming multitudes from poverty and hopelessness into an ocean of prosperity, built on the aspirational forward-looking pedestal of global influence and respect.

In addition, the articulation of India's multiple self-identities draw from three specific impulses.

6 LOOKING AHEAD

Coal Will Fuel Growth in the Foreseeable Future

There is a solemn recognition that India's developmental needs have to be fuelled by conventional energy. And concomitantly, the pressures of affordability and access will dominate the considerations of emission morality and leadership responsibility. This will continue to be seen, even as a balance will be sought between the two.

Cheap energy generated by fossil fuels has been the bedrock upon which every industrialised nation – from the United Kingdom, the United States and Germany in the nineteenth century, to most recently China in the twentieth century – have achieved development goals. There exists a strong inter-linkage between energy access and development. Close to 307 million or a quarter of India's population have no access to electricity, presenting perhaps the most daunting energy access challenge in the history of mankind. Yet this represents a vastly improved figure from as recent as two decades ago, when close to half of the population had no access to electricity, the point at which India embarked on the path of open market and liberalisation.

Today, when after almost two and a half decades of painful reform, as India stands at the cusp of lifting millions of its citizens out of poverty into prosperity, albeit on the back of industrialisation fuelled by cheap non-renewable fossil fuels, to expect it to put the process on hold until a cost-competitive renewable option is available is not realistic. Yet India recognises that the biggest source of energy is energy efficiency – a gross financial saving exceeding 20% by way of smarter energy management – and this margin play afforded within the current energy set-up is what India will move fast to leverage. This is not a roadblock to do more and is not colliding with an ambition to develop a green economy. This mix of *energy security* and *industrial identities* will remain key to India for many years to come.

Enterprise Has to Complement It

Arising from the first is the second impulse where India recognises the long-term non-viability of fossil-based energy and encourages a massive national programme of energy entrepreneurship. Represented in the *entrepreneurial identity*, this opportunity to leverage technology lies in replacing and upgrading existing energy-intensive systems within all sectors, including agriculture and industry.

India has punched far above its weight in this regard. For instance, India's expenditure on renewable energy (solar photovoltaic and wind energy) has far exceeded more prosperous counterparts. Indians on an average spent about 1.5 times more than the average Chinese, about two to four times more than the average Japanese and two times more than the average American on solar energy, for instance (Saran and Sharan 2015).

Even as the federal government has given several policy sops, including tax breaks, duty waivers and cheaper loans to promote green energy, governments at the state level too are aggressively following suit. The government in the western state of Maharashtra, for instance, recently introduced a 'New and Renewable Energy Policy' with a target of generation of 14,400 MW through renewables alone.

Indeed, the narrative of green enterprise or a green economy is closely tied to the imperative of job creation. The young in particular are driving this narrative. So is media reportage, which sees India as both an emerging nation at times, which has the will and the wherewithal to develop technology and green industry, and, at other times, sees it as a developing country, which must solicit and look to benefit from both financial and technology flows from the Global North (which has a historical responsibility to catalyse such flows).

India Is Ready for a Seat at the Global High Table

There is wide recognition of the transforming role of India in the world, from a developing nation to a nation of rapid economic growth and leadership on matters of global governance. The *emerging nation self-identity* builds on this aspirational crest and finds strong resonance in media reportage, as well as expert interviews. India's large size and energy needs give it veritable veto in climate negotiations. Yet its approach has not been unilateral. Instead, since the very beginning, India has sought to build consensus on negotiating positions and outcomes, particularly with China.

Today as China's energy priorities have shifted (evident from its 2014 bilateral deal with the United States on climate change on the margins of the Asia Pacific Economic Cooperation Summit) as it has become a middle-income nation, India has to look to forge its position within the South–South construct but not necessarily dependant on Chinese support.

India's official position at the climate negotiations has historically been built around the *rural* and *developing country* narratives. Along with

Group of 77 countries, India has made many forceful representations positing that the developed world has a historical responsibility to cut emissions, and simultaneously a responsibility to help the developing world build mitigating and adapting capacities and capabilities. This position has found resonance within the Indian media, particularly during and after the COP negotiations even as at other times the media narrative is built around an aspirational worldview.

It is perhaps an unfortunate irony that just as India readies for industrialisation, and as it readies to lift the living standards of its teeming millions, multilateral action agenda is being propelled towards carbon emission cuts, ostensibly to make the world a better more prosperous place! Yet, with its vast coastline and glacial cover, India is among the most vulnerable to climate change, and the consequences of climate inaction will be more pronounced on it than most. This dilemma calls for an approach that balances India's development needs with its climate responsibility. And this can come only through decisive executive action and leadership, away from the current narrative of inefficient entitlements and offsets.

This requires larger recognition of the problem. It also necessitates finding meaningful meeting grounds, and working for the larger common cause. Away from the pressures of coalition politics, the progressive observed pre-COP sentiment, and research that posits such affirmative action as advantage, the current government in India is perfectly situated to make a leap of faith. It must build on the *industrial, emerging* and *entrepreneurship self-identities.* Allowing the organic articulation of progressive identities, without infusing the minimalist agenda for the sake of political expediency, will not only align better to the aspirational world view of India's youth, but will also find natural resonance with them. Such harnessing of the youth's natural agency will diminish the possibility of any political backlash, the very reason minimalist agendas are being pursued in the first place. Doing so would afford the government sufficient leeway to negotiate and conclude global climate agreements that better marry its domestic imperatives with its ethical responsibility towards the greater good.

For a government that draws from India's great cultural legacy, in its aspiration to build a modern and progressive nation at the forefront of humanity, *Vasudhaiva Kutumbakam* from the *Atharva Veda* dating to 3000 BC affords every validation. For the world, the phrase explains, is one family, where the greater welfare of all as against the narrow regard for self, is the ultimate touchstone. With India taking a leading role in the campaign for climate justice, motivated as much by its own national interests as its aspiration to be the reasoned and tempered voice of the developing and emerging world, it will have to tread carefully, as these identities become less fungible, in contested global spaces.

For India, discovering the finite shape of this amorphous set of identities is its 'climate identity'.

BIBLIOGRAPHY

Adger, W., Barnet, J., Chapin, F., & Ellemor H.(2011). *This must be the place: underrepresentation of identity and meaning in climate change decision-making.* http://www.lter.uaf.edu/pdf/1553_Adger_Barnett_2011.pdf. Accessed 5 August 2015.

Agarwal, A., & Narain, S. (1991). Global warming in an unequal world: A case of environmental colonialism. http://cseindia.org/challenge_balance/readings/GlobalWarming%20Book.pdf. Accessed 5 January 2015.

Ahn, S. J., & Graczyk, D. (2012). *Understanding energy challenges in India—Policies, players and issues.* Partner Country Series. http://www.iea.org/publications/freepublications/publication/India_study_FINAL_WEB.pdf. Accessed 4 January 2014.

Akhtar, M. J., Ali, A. A., & Akhtar, S. (2010). The role of Vernacular press in subcontinent during the British rule: A study of perceptions. *Pakistan Journal of Social Sciences (PJSS), 30*(1), 71–84.

Alkire, S. (2002). Dimensions of human development. http://www.unicef.org/socialpolicy/files/Dimensions_of_Human_Development.pdf. Accessed 5 August 2015.

Ananthapadmanabhan, G., Srinivas, K., & Gopal, V. (2007). *Hiding behind the poor. Greenpeace Report on Climate Injustice.* http://www.greenpeace.org/india/Global/india/report/2007/11/hiding-behind-the-poor.pdf. Accessed 9 May 2014.

Bali. (2002). Bali principles of climate justice. http://www.ejnet.org/ej/bali.pdf. Accessed 10 September 2014.

Bandura, A. (1994). Self-efficacy. In R. J. Corsini (Ed.), *Encyclopaedia of psychology* (pp. 368–369). 2nd edition. New York: Wiley.

Banga, R. (2005). *Critical issues in India's service led growth.* http://icrier.org/pdf/WP171.pdf. Accessed 9 June 2014.

Bhoyar, S. P., Dusad, S., Shrivastava, R., Mishra, S., Gupta, N., & Rao, A. B. (2014). Understanding the impact of lifestyle on individual carbon-footprint. *Procedia-Social and Behavioral Sciences, 133,* 47–60.

Billett, S. (2009). Dividing climate change: Global warming in the Indian mass media. http://sciencepolicy.colorado.edu/students/envs_4800/billett_2009.pdf. Accessed 15 June 2014.

Bureau of Energy Efficiency. (2008). *Annual report 2007–08, Ministry of Power, Government of India.* http://beeindia.in/about_bee/documents/annualreports/2007-08E.pdf. Accessed 25 July 2014.

Census of India. (2011). Ministry of home affairs. *Government of India.* http://censusindia.gov.in/. Accessed 15 July 2014.

Central Electricity Authority, Government of India. (2015). Power sector—Executive summary. http://www.cea.nic.in. Accessed 2 April 2015.

Chakravarty, S., & Ramana, M. V. (2012). The hiding behind the poor debate. In N. Dubash (Ed.), *Handbook of climate change and India: Development, politics and governance* (pp. 218–226). London: Earthscan.

Colomer, J. (1995). *Leadership games in collective action.* http://rss.sagepub.com/content/7/2/225.short. Accessed 13 October 2014.

Confederation of Indian Industries. (2013). *Climate change.* http://www.cii.in/Sectors.aspx?enc=prvePUj2bdMtgTmvPwvisYH+5EnGjyGXO9hLECvTuNvtoshNFjXWF9pWAvpZBDgh. Accessed 5 June 2014.

Consolidated Energy Consultants Limited. (2011). *Assessment of investment climate for wind power development in India for Indian Renewable Development Agency* (IREDA), New Delhi. Bhopal: Consolidated Energy Consultants Limited. http://ireda.gov.in/writereaddata/Assessment.pdf. Accessed 12 August 2014.

Dasgupta, C. (2012). Present at the creation: The making of the UN framework convention on climate change. In N. Dubash (Ed.), *Handbook of climate change and India: Development, politics and governance.* New Delhi: Oxford University Press.

Datt, D., Norohna, L., Srivastava, N., & Sridharan, P. V. (2009). Resource federalism in India: The case of minerals. *Economic and Political Weekly, XLIV*(8), 51–59.

Deaton, A., & Dreze, J. (2002). Poverty and Inequality in India: A re-examination. *Economic and Political Weekly,* 3729–3748.

Deccan Chronicle. (2015). *India should guide the world on fighting climate change, says Narendra Modi.* http://www.deccanchronicle.com/150406/nation-current-affairs/article/india-should-have-taken-lead-climate-change-modi-two-day. Accessed 7 April 2015.

Dubash, N. (2009). Copenhagen: Climate of mistrust. *Economic and Political Weekly*, *XLIV*(52), 8–11.

Dubash, N. (ed.). (2012). *Handbook of climate change and India: Development, politics and governance.* London & New York: Earthscan.

Dubash, N. (2013). The politics of climate change in India: Narratives of equity and co-benefits. *WIREs Climate Change*, *4*(3), 191–201.

Dubash, N., & Rajan, S. (2001). *The politics of power sector reform in India.* http://pdf.wri.org/power_politics/india.pdf. Accessed 3 January 2015.

Felipe, J., Kumar, U., &Abdon, A. (2010). Exports, capabilities, and industrial policy in India. Agricultural sector overview. http://www.ficci-b2b.com/sector-overview-pdf/Sector-agri.pdf. Accessed 31 March 2014.

FICCI-KPMG. (2013). *The power of a billion: Realizing the Indian dream.* http://www.ficci.com/spdocument/20217/FICCI-KPMG-Report-13-FRAMES.pdf. Accessed 11 July 2014.

Finance Commission of India. (2015). *14th Finance commission report.* New-Delhi, India: Government of India.

Firstpost. (2015). India will push for polluter pays policy at Paris climate change meet, says Prakash Javadekar—Firstpost. *Firstpost.* http://www.firstpost.com/india/india-will-seek-climate-justice-for-the-billions-of-poor-of-this-world-at-paris-climate-change-meet-2517138.html. Accessed 27 March 2016.

Ford, D., & Urban, H. (1963). *Systems of psychotherapy: A comparative study.* New York: John Wiley & Sons Inc. Press.

Frohlich, N., Oppenheimer, J., &Young, O. (1971). *Political leadership and collective goods.* Princeton: Princeton University Press. Cited from: Esteban, J., & E. Hauk (2008) Leadership in Collective Action, p. 2.

Gandhi, I. (2008). *Of man and his environment.* India: Abhinav Publications, pp. 1–23.

Gaye, A. (2007). *Access to energy and human development.* http://hdr.undp.org/sites/default/files/gaye_amie.pdf. Accessed 20 July 2014.

Ghosh, P. (2012). Climate change debate: The rationale of India's position. In N. K. Dubash (Ed.), *Handbook of climate change and India: Development, politics, and governance.* London, Oxford: Oxford University Press.

Government of India, Forest Conservation Act. (1980). http://www.moef.nic.in/legis/forest/forest2.html. Accessed 31 March 2014.

Grover, A. (2003). *Electricity Act 2003: Opportunities and threats—Crisil young thought leaders.* http://www.crisil.com/youngthoughtleader/winners/07-Grover-NM.PDF. Accessed 31 March 2014.

Hallding, K., Olsson, M., Atteridge, A., Carson, M., Vihma, A., & Roman, M. (2011). *Report preview: Together alone: Brazil, South Africa, India, China (BASIC) and the climate change conundrum.* Stockholm: Stockholm Environment Institute.

Hanrath, J., & Leggewie, C. (2013). *Revolution 2.0? The role of digital media in political mobilisation and protest.* https://www.academia.edu/10198612/Revolution_2.0_The_Role_of_Digital_Media_in_Political_Mobilisation_and_Protest. Accessed 9 May 2014.

HDRO Outreach. (2015). What is human development? http://hdr.undp.org/en/content/what-human-development. Accessed 5 August 2015.

Helms, L. (2014). Global political leadership in the twenty-first century: Problems and prospects. *Contemporary Politics, 20*(3), 261–277.

Hilgartner S., & Bosk, C. (1988). *Rise and fall of social problems: A public arenas model*http://www.unc.edu/~fbaum/teaching/PLSC_SOC_497_SP_2008/Hilgartner_Bosk_AJS_1988.pdf. Accessed 31 July 2014.

Hoffman, A. (2004). *Climate change strategy: The business logic behind voluntary greenhouse gas reductions.* Ross School of Business Paper No. 905.

Hoffman, A., & Woody, J. (2008). *Climate change: What's your business strategy?* Harvard: Harvard Business Press.

Hurrell, A., & Sengupta, S. (2012). Emerging powers, North–South relations and global climate politics. *International Affairs, 88*(3), 463–484.

India Disaster Knowledge Network. (2009). *Disaster profile.* http://www.saarc-sadkn.org/countries/india/disaster_profile.aspx. Accessed 20 July 2014.

India stats. (n.d.). *Indian market economy.* http://www.indiastat.com/economy/8/stats.aspx. Accessed 1 July 2014.

International Energy Agency. (2009). *World energy outlook 2009.* France: International Energy Agency (IEA).

International Energy Agency. (2011). *World energy outlook 2011.* France: International Energy Agency (IEA). http://www.iea.org/publications/freepublications/publication/weo2011_web.pdf. Accessed 16 July 2014.

Internet and Mobile Association of India and IMRB International. (2014). Internet in India 2014. New-Delhi, India.

Internet Live Stats. (2015). Elaboration of Internet users' data by Internet & Mobile Association of India (IAMAI). International Telecommunication Union (ITU), World Bank, and United Nations Population Division. http://www.internetlivestats.com/. Accessed 7 August 2014.

Jain, S. (2006) Political economy of the electricity subsidy: evidence from Punjab, *Economic and Political Weekly, 41,* 4072–4080

Jain, S. (2015). India: Multiple media explosions. In K. Nordenstreng, & D. K. Thussu (Eds.), *Mapping BRICS media* (pp. 151–165). New York: Routledge.

Jaiswal, S., Gupta, A., & Mankad, M. (2010). India's not waiting. http://switchboard.nrdc.org/blogs/ajaiswal/indias_not_waiting.html. Accessed 20 December 2014.

Javadekar, P. (2014). Statement made at the 20th Conference of Parties (COP-20) in Lima. http://envfor.nic.in/content/statement-hon%E2%80%99ble-

minister-high-level-segment-unfccc-cop-20-december-9-2014. Accessed 27 March 2016.
Jones, C. (2015). Rethinking the legacy of President Obama. http://www.huffingtonpost.com/clarence-b-jones/rethinking-the-legacy-of-president-obama_b_6923550.html?ir=India&adsSiteOverride=in. Accessed 1 April 2015.
Kasa, S., Gullberg, A., & Heggelund, G. (2008). The Group of 77 in the international climate negotiations: Recent developments and future directions. *International Environmental Agreements: Politics, Law and Economics, 8*(2), 113–127.
Khare, V., Nema, S., & Baredar, P. (2013). Status of solar wind renewable energy in India. *Renewable and Sustainable Energy Reviews, 27*, 1–10.
King, D. (2002). The changing shape of leadership. *Educational Leadership, 59*(8), 61–63.
Krishna, S. M. (2012). At the Interactive Debate on '*Lasting Peace through Joint Global Governance*'. At the Preparatory Ministerial Meeting of the Non-Aligned Movement Tehran. Iran on 28 August 2012. http://www.un.int/india/nam/2012eam.nam.pdf. Accessed 10 September 2014.
Kumar, M. (2013, June, Summer Issue). Comparison of science coverage in Hindi and English newspapers of India: A content analysis approach. *Global Media Journal, 4*(1), 8–11.
Kumar, S., & Sarma, V. V. S. (2015). Performance and Challenges of Newspapers in India: A Case Study on English versus Vernacular Dailies in India. http://www.aims-international.org/aims12/12A-CD/PDF/K740-final.pdf. Accessed 15 April 2015
Lakoff, G. (1996). *Moral politics: What conservatives know that liberals don't author: George Lakoff.* University of Chicago Press, p. 421.
Lankala, S. (2006). Mediated nationalisms and 'Islamic terror': The articulation of religious and post-colonial secular nationalism in India. *Westminster Papers in Communication and Culture, 3*(2), 86–102.
Leiserowitz, A., & Thaker, J. (2012). *Climate change in the Indian mind.* http://environment.yale.edu/climate-communication/files/Climate-Change-Indian-Mind.pdf. Accessed 10 September 2014.
Levinas, E. (1963). The trace of the other. *Tijdschrigtvoor Philosophie* (trans: A. Lingis), pp. 605–623.
Lok Sabha (2013). Unstarred Question No. 1227. Answered on 05. March 2013. Government of India.
Macnamara, J. (2005). Media content analysis: Its uses, benefits and best practice methodology. *Asia Pacific Public Relations Journal, 6*(1), 1–34.
Malhotra, I. (2008). *Changing face of Indian media.* http://www.mediamimansa.com/6th%20issue/6eng_67-72.pdf. Accessed 13 October 2014.

Masciulli, J., Molchanov, M. A., & Knight, W. A. (2009). Political leadership in context. In J. Masciulli, M. A. Molchanov & W. A. Knight (Eds.), *The Ashgate Research Companion to Political Leadership* (pp. 3–27). London: Routledge.

Mashinhur, R. (2008). Interactive option in online newspapers of Bangladesh. *Pakistan Journal of Social Sciences*, 5(6), 620–624.

Maslow, A. H., Frager, R., & Cox, R. (1970). *Motivation and personality*. Vol. 2. J. Fadiman, & C. McReynolds (Eds.). New York: Harper & Row.

Maturana, H., (1988). Ontology of Observing: The Biological Foundations of Self-Consciousness and the Physical Domain of Existence. http://ada.evergreen.edu/~arunc/texts/cybernetics/oo/old/oo.pdf. Accessed 22 September 2016.

Mazzoleni, G. (2003). The media and the growth of neo-populism in contemporary democracies. In M. L. Gat, G. Mazzoleni & J. Stewart (Eds.), *The Media and Neo-Populism: A Contemporary Comparative Analysis* (pp. 1–21). Santa Barbara, CA: Praeger.

Mburia, R. (2015). *Africa climate change policy: An adaptation and development challenge in a dangerous world*. http://www.fao.org/fsnforum/sites/default/files/resources/AFRICA%20CLIMATE%20CHANGE%20POLICY-CEI.pdf. Accessed 4 August 2015.

McClelland, D. (1985). How motives, skills, and values determine what people do. *American Psychologist*, 40(7), 812.

McKibben, B. (2012). Global warming's terrifying new math. *Rolling Stone, 19*, 43. http://www.rollingstone.com/politics/news/global-warmings-terrifying-new-math-20120719. Accessed 18 August 2014.

McManus, P. A. (2000). Beyond Kyoto? Media representation of an environmental issue. *Australian Geographical Studies*, 38(3), 306–319.

Media Research Users Council. (2014). Indian readership survey 2014. Mumbai, India. http://mruc.net/sites/default/files/IRS%202014%20Topline%20Findings_0.pdf. Accessed 1 March 2015.

Mehrishi, M. (2012). *Speech at 18th Conference of Parties (COP 18)*.http://unfccc.int/resource/docs/cop18_cmp8_hl_statements/Statement%20by%20India%20(COP%2018).pdf. Accessed 5 January 2014.

Miller, A. (1961, November 26). As quoted in The Observer. London, UK.

Mineral Concessions Rules, Government of India. (1960). http://www.coalindia.in/RTI_crchp3.aspx. Accessed 12 July 2013.

Ministry of Coal. (2011). The year 2010–11 at a glance, Government of India. http://www.coal.nic.in/annrep1011.pdf. Accessed 12 July 2013.

Ministry of Coal. (n.d.). Note on legislation, government of India. http://www.coal.nic.in/policy/legislation.pdf. Accessed 12 July 2013.

Ministry of Commerce and Industry. (2014). Foreign trade performance of India. annual report 2013–14. Directorate General of Commercial Intelligence and

Statistics. Government of India. http://www.dgciskol.nic.in/annualreport/book_3e.pdf. Accessed 15 March 2015.

Ministry of Environment and Forest Government of India. (2010). Letter to the executive secretary, United Nations framework convention on climate change, Germany. https://unfccc.int/files/meetings/cop_15/copenhagen_accord/application/pdf/indiacphaccord_app2.pdf. Accessed 5 January 2014.

Ministry of Environment, Forests and Climate Change. (2011). Government of India. *Sustainable Development in India: Stocktaking in the Run-Up to Rio+20*, Government of India.

Ministry of Environment, Forests and Climate Change. (2012). *Second national communication to the United Nations framework convention on climate change*. Government of India. http://unfccc.int/resource/docs/natc/indnc2.pdf. Accessed 10 September 2014.

Ministry of External Affairs. (2013). Fourth India-U.S.A strategic dialogue. Joint Statement–MEA. http://mea.gov.in/bilateral-documents.htm?dtl/21872/Joint+Statement+on+the+Fourth+IndiaUS+Strategic+Dialogue. Accessed 15 December 2013.

Ministry of External Affairs, Government of India. (2007). *PM's intervention on climate change at the Heiligendamm meeting on 8 June 2007*. New-Delhi: Government of India.

Ministry of Home Affairs. (2010). Naxal management division. http://mha.nic.in/Naxal. Accessed 4 September 2014.

Ministry of Information and Broadcasting. (2013). Annual report of the registrar of newspapers for India under the Press and Registration of Books Act, 1867. 51 vols. New Delhi: Government of India, 7–28–56.

Ministry of Information and Broadcasting. (2014). *Press in India 2013–14*. http://www.rni.nic.in/pin1314.pdf. Accessed 1 January 2015.

Ministry of New and Renewable Energy. (2010). *Jawaharlal Nehru solar mission—Towards building a solar India*. Mission Document. Government of India.

Ministry of Petroleum and Natural Gas. (2013). *Snapshot of India's Oil and Gas data. Petroleum and planning analysis cell*. New Delhi: Government of India.

Ministry of Power. (2006). Tariff policy. The Gazette of India, PART I—Section 1. Published by Authority: Ministry of Power: Government of India.

Ministry of Power. (2013). *Indian electricity scenario*. Government of India. http://powermin.nic.in/indian_electricity_scenario/national_electricity_policy.htm. Accessed 10 December 2013.

Ministry of Power. (2013a). Measures to reduce AT&C losses. http://pib.nic.in/newsite/PrintRelease.aspx?relid=93530. Accessed 15 March 2016.

Ministry of Power. (2016). Total installed capacity. http://powermin.nic.in/power-sector-glance-all-india. Accessed 15 April 2016.

Ministry of Power, Government of India. (2005). *National electricity policy.* http://www.powermin.nic.in/whats_new/national_electricity_policy.htm. Accessed 7 June 2014.

Ministry of Statistics and Program Implementation. (2013). *Energy statistics—2013.* New-Delhi, India.

Ministry of Water Resources. (2012). Government of India. http://mowr.gov.in/writereaddata/linkimages/DraftNWP2012_English9353289094.pdf. Accessed 12 July 2014.

Modi, N. (2015a). Narendra Modi's speech at UNESCO in full. http://blogs.wsj.com/indiarealtime/2015/04/10/narendra-modis-speech-at-unesco-in-full/. Accessed 11 April 2015.

Modi, N. (2015b). *Statement by Prime Minster at COP 21 Paris, November 2015.* Paris: UNFCCC. https://unfccc.int/files/meetings/paris_nov_2015/application/pdf/cop21cmp11_leaders_event_india.pdf. Accessed 27 March 2016.

Natarajan, J. (2011). *Speech at 17th Conference of Parties (COP 17) in Durban.* http://www.moef.nic.in/downloads/public-information/MEF%20statement_HLS_CoP17_Dec-07.pdf. Accessed 5 January 2014.

Natarajan, J. (2013). Statement made at 19th Conference of Parties (COP-19). Warsaw: UNFCCC. http://www.ieee.es/Galerias/fichero/Varios/Cumbre_CambioClimatico_Varsovia_Nov2013.pdf Accessed 27 March 2016.

National Book Trust, India and National Council of Applied Economic Research. (2010). *Indian youth demographics and readership: Results from national youth readership survey.* New Delhi: NCAER, pp. 9–56.

National Electricity Policy. (2005). Ministry of power, government of India. http://www.powermin.nic.in/whats_new/national_electricity_policy.htm. Accessed 12 July 2014.

National Sample Survey. (2008–2009). Ministry of statistics and programme implementation. Government of India.

National Sample Survey. (2011). Ministry of statistics and programme implementation. Government of India.

Nin, A. (1961). *Seduction of the Minotaur.* Swallow Press. p. 1. Ohio: Ohio University Press.

Norris, P. (2000). *A virtuous circle: Political communications in post-industrial societies.* Cambridge: Cambridge University Press.

O'Neill, J. (2001). Building better global economic BRICs. http://www.goldmansachs.com/our-thinking/archive/archive-pdfs/build-better-brics.pdf. Accessed 12 December 2013.

Pal, P., & Ghosh, J. (2007). Inequality in India: A survey of recent trends. In *Economic and Social Affairs.* Working Paper 45. New York: Department of Economics and Social Affairs.

Pew Charitable Trust & B.N.E. Finance. (2010). Who's winning the clean energy race. In *Growth, competition and opportunity in the world's largest economies.*

Washington, DC: The Pew Charitable Trust. http://www.pewtrusts.org/~/media/legacy/uploadedfiles/peg/publications/report/g20reportlowresfi nalpdf.pdf. Accessed 10 September 2014.

Planning Commission. (2002). *National human development report 2001.* New Delhi: Government of India.

Planning Commission. (2011a). Expert group on low carbon strategies for inclusive growth (2011). *Low carbon strategies for inclusive growth: An interim report.* New Delhi: Government of India.

Planning Commission. (2011b). Power and energy. http://planningcommission. nic.in/sectors/index.php?sectors=energy. Accessed 10 August 2014.

Planning Commission of Government of India. (2012). Databook on the Indian Economy. http://planningcommission.nic.in/data/datatable/0814/comp_databook.pdf. Accessed 15 July 2014.

Power Finance Corporation Limited. (2011). *Draft shelf prospectus.* http://www.sebi.gov.in/dp/powerfindebt.pdf. Accessed 10 September 2014.

Press Trust of India. (2012). *BRICS summit: Pacts signed to promote trade in local currencies.* http://www.dnaindia.com/india/report-brics-summit-pacts-signed-to-promote-trade-in-local-currencies-1668990. Accessed 7 June 2014.

Prime Minister's Office. (2013). Government of India. http://pmindia.nic.in/cmp.pdf. Accessed 3 December 2013.

Public Finance Corporation Ltd. (2011). Annual report 2010–2011. http://www.pfcindia.com/writereaddata/userfiles/file/Annual%20reports/ann_rpt1011.pdf. Accessed 10 August 2014.

Puryear, S. (2008). Leibniz on concepts and their relation to the senses. In D. Perler & M. Wild (Eds.), *Sehen und Begreifen: Wahrnehmungstheorien in der Frühen Neuzeit* (p. 2). New York: de Grunter.

Putnam, R. (1998, Summer). Diplomacy and domestic politics. The logic of two way games. *International Organisation, 42*(3), 427–460.

Qi, X. (2011). The rise of BASIC in UN climate change negotiations. *South African Journal of International Affairs, 18*(3), 295–318.

Radcliffe, S. (2005). Development and geography: Towards a post colonial development geography? *Progress in Human Geography, 29*(3), 291–298.

Raghuvanshi, S., Chandra, A., & Raghav, A. (2006). Carbon dioxide emissions from coal based power generation in India. *Energy Conservation and Management. Elsevier, 47,* 427–441.

Rajasekhar, D., & Sahu, G. (2006). The growing rural -urban disparity: Some issues. http://www.isec.ac.in/wp.html#top. Accessed 9 May 2014.

Rajiv Gandhi Grameen Vidyutkaran Yojna. (2012). Government of India. http://www.rggvy.gov.in/rggvy/rggvyportal/plgsheet_frame3.jsp. Accessed 10 December 2013.

Ram, N. (2013). 'The changing role of news media in contemporary India', at the 72nd Indian History Congress, Punjab University (2011), Patiala. p .7.

Ramachandran, V. K. (1996). On Kerala's development achievements. In J. Dreze, & A. Sen (Eds.), *Indian development: Selected regional perspectives* (p. 206.). Oxford: Clarendon Press.

Ramesh, J. (2009). *Jairam Ramesh statement on Copenhagen Accord in Rajya Sabha*. http://www.thehindu.com/news/national/jairam-ramesh-statement-on-copenhagen-accord-in-rajya-sabha/article69893.ece. Accessed 15 March 2014.

Ramesh, J. (2010a). *Letter to parliamentarians on the cancun agreement*. http://www.sanctuaryasia.com/index.php?view=article&catid=122%3Aclimate-change&id=3929%3Aletter-from-jairam-ramesh-on-the-cancun-agreement&option=com_content&Itemid=289. Accessed 5 January 2014.

Ramesh, J. (2010b). *Speech at conference of parties to the UNFCCC (COP-16)*. http://admin.indiaenvironmentportal.org.in/files/speech-cancun.pdf. Accessed 5 January 2014.

Rao, N., Sant, G., & Rajan, S. C. (2009). *An overview of Indian energy trends: Low carbon growth and development challenges*. Pune: Prayas Energy Group.

Rejikumar, R. (2005). National electricity policy and plan: A critical examination. *Economic and Political Weekly, XL*(20), 2028–2032.

Rogers, C., Kell, B., & McNeil, H. (1948). The role of self-understanding in the prediction of behaviour. *Journal of Consulting Psychology, 12*, 174–186.

Sant, G., & Gambhir, A. (2012). Energy, development and climate change. In *Handbook of Climate Change and India*. Oxford University Press.

Saran, S., & Jones, B. (2015). An 'India Exception' and India-US partnership on climate change. http://orfonline.org/cms/export/orfonline/modules/issuebrief/attachments/ORF_Issue_brief_85_1421127922603.pdf. Accessed 25 February 2015.

Saran, S., & Sharan, V. (2012a). Giving BRICS a non-western vision. http://www.thehindu.com/opinion/op-ed/article2889838.ece?mstac=0. Accessed 30 January 2014.

Saran, S., & Sharan, V. (2012b). Identity and energy access in India—Setting contexts for Rio+20. http://bookstore.teri.res.in/docs/newsletters/ESI%207(1)%20Jan-Mar%202012.pdf. Accessed 7 February 2014.

Saran, S., & Sharan, V. (2013). More than just a catchy acronym: Six reasons why BRICS matters. http://www.globaltimes.cn/content/754826.shtml. Accessed 5 February 2014.

Saran, S., & Sharan, V. (2015). Indian leadership on climate change: Punching above its weight. *Brooking blog*. http://www.brookings.edu/blogs/planetpolicy/posts/2015/05/05-indian-leadership-climate-change-saran-sharan. Accessed 10 May 2015.

Sarukkai, S. (1997). The other in anthropology and philosophy. *Economic and Political Weekly, 32*(24), 1406–1409.

Sathaye, J., Shukla, P., & Ravindranath, N. (2006). Climate change, sustainable development and India: Global and national concerns. *Current Science-Bangalore*, *90*(3), 314–325.

Savaresi, A. (2016). The Paris agreement: A new beginning? *Journal of Energy & Natural Resources Law*, *34*(1), 16–26.

Sengupta, S. (2012). International climate negotiations and India's role. In N. K. Dubash (Ed.), *Handbook of climate change and India*. New Delhi: Oxford University Press.

Sethi, N. (2014, September 18). Javadekar to lead at Ban Ki-moon's climate summit in New York. http://www.business-standard.com/article/economy-policy/javadekar-to-lead-at-ban-ki-moon-s-climate-summit-in-new-york-114091800023_1.html. Accessed 10 November 2014.

Sharan, V., & Kumar, P. (2013). South–South cooperation: A survey of recent trends. http://orfonline.org/cms/export/orfonline/modules/issuebrief/attachments/issuebrief65_1388636496297.pdf. Accessed 20 February 2014.

Shukla, N., & Dixit, D. (2007). *Socio-economic disparities in India*. New Delhi: Anmol Publications.

Shukla, P. R., & Chaturvedi, V. (2013). Sustainable energy transformations in India under climate policy. *Sustainable Development*, *21*(1), 48–59.

Sims, R., Schock, R., Adegbululgbe, A., Fenhann, J., Konstantinaviciute, I., Moomaw, W., et al. (2007). Energy supply. In B. Metz, O. R. Davidson, P. R. Bosch, R. Dave, & L. A. Meyer (Eds.), *Climate change 2007: Mitigation. Contribution of working group III to the fourth assessment report of the intergovernmental panel on climate change*. Cambridge: Cambridge University Press.

Singh, H. (1994). Constitutional base for Panchayati Raj in India: The 73rd Amendment Act. *Asian Survey*, 818–827.

Sinha, Y. (2002). *Speech at the World Summit on Sustainable Development*. http://www.un.org/events/wssd/statements/indiaE.htm. Accessed 5 January 2014.

Smith, R. (2004). *Strategic planning for public relations*. Routledge, p. 296.

Sprinz, D. F., & Weiß, M. (2001). Domestic politics and global climate policy. In U. Luterbacher & D. F. Sprinz (Eds.), *International relations and global climate change* (pp. 67–94). Massachusetts: MIT.

Srivas, A. (2015). Fighting climate change needs climate justice, Says India. *The Wire*. http://thewire.in/2015/09/26/international-partnerships-must-for-fighting-climate-change-says-modi-11668/. Accessed 27 March 2016.

Stuenkel, O. (2010). *Responding to global development challenges: Views from Brazil and India*. https://www.die-gdi.de/uploads/media/DP_11.2010.pdf. Accessed 9 May 2014.

Subramanian, N. (2007). Populism in India. *SAIS Review of International Affairs*, *27*(1), Winter–Spring, 81–91.

Sundaram, K., & Tendulkar, S. (2003). Poverty in India in the 1990s: An analysis of Changes in 15 Major States. *Economic and Political Weekly, XXXVIII*(14), 1385–1393.

The Group of 77 at the United Nations. (n.d.). Introduction. http://www.g77.org/doc/CPA-intro.htm. Accessed 28 August 2014.

The New York Times. (2008). *Barack Obama's remarks in St. Paul.* http://www.nytimes.com/2008/06/03/us/politics/03text-obama.html?pagewanted=all&_r=0. Accessed 29 January 2015.

The Pew Research Center. (2008). *Media credibility. US politics & policy.* http://www.people-press.org/2008/08/17/media-credibility/. Accessed 2 March 2015.

The White House. (2015).Climate change and president Obama's action. United States of America. https://www.whitehouse.gov/climate-chang. Accessed 2 April 2015.

Trumbo, C. W., & Shanahan, J. (2000). Social research on climate change: Where we have been, where we are, and where we might go. *Public Understanding of Science, 9,* 199–204.

United Nations. (1993). *Report of the United nations conference on environment and development.* Vol. III. Statements made by head of states or Government at the Summit Segment of the Conference. New York: United Nations, pp. 1–3.

United Nations Development Programme. (2005). *Energizing the millennium development goals: A guide to energy's role in reducing poverty.* New York: United Nations Development Programme, p. 14.

United Nations Environment Programme. (1992). Rio Declaration on Environment and Development, Agenda 21, the statement on forest principles, United Nations Framework Convention on Climate Change, and United Nations Convention on Biodiversity. Rio De Janeiro: United Nations. http://www.un.org/geninfo/bp/enviro.html Accessed 27 March 2016.

United Nations Framework Convention on Climate Change. (1993). *Intergovernmental negotiating committee for a framework convention on climate change.* http://unfccc.int/resource/docs/a/50add1.pdf. Accessed 6 August 2014.

United Nations Framework Convention on Climate Change. (1997). Kyoto protocol. http://unfccc.int/kyoto_protocol/items/2830.php. Accessed 10 October 2014.

United Nations Framework Convention on Climate Change. (2009a). Copenhagen accord. http://unfccc.int/meetings/copenhagen_dec_2009/items/5262.php. Accessed 10 September 2014.

United Nations Framework Convention on Climate Change. (2009b). Decision -/CP.15. http://unfccc.int/files/meetings/cop_15/application/pdf/cop15_cph_auv.pdf. Accessed 10 September 2014.

United Nations Framework Convention on Climate Change. (2010). *Report of the conference of the parties on its sixteenth session, held in Cancun from 29 November to 10 December 2010.* http://unfccc.int/resource/docs/2010/cop16/eng/07a01.pdf. Accessed 10 September 2014.

United Nations Framework Convention on Climate Change. (2011). *Durban: Towards full implementation of the UN climate change convention.* http://unfccc.int/key_steps/durban_outcomes/items/6825.php. Accessed 10 September 2014.

United Nations Framework Convention on Climate Change. (2012). Conference of parties 18, Doha, Qatar. http://unfccc.int/meetings/doha_nov_2012/meeting/6815.php. Accessed 10 September 2014.

United Nations Framework Convention on Climate Change. (2013). Report of the conference of the parties on its nineteenth session, held in Warsaw from 11 to 23 November 2013. http://unfccc.int/resource/docs/2013/cop19/eng/10a01.pdf. Accessed 10 September 2014.

United Nations Framework Convention on Climate Change. (2015). Paris agreement. https://unfccc.int/resource/docs/2015/cop21/eng/l09r01.pcf. Accessed 24 June 2016.

Upadhyay, D., & Mishra, M. (2013). Emerging dynamics of climate change: Post—Doha climate gateway. *Indian Foreign Affairs Journal, 8*(2), 204–212.

Vihma, A., Mulugetta, Y., & Karlsson-Vinkhuyzen, S. (2011). Negotiating solidarity? The G-77 through the prism of climate change negotiations. *Global Change, Peace & Security, Formerly Pacifica Review: Peace, Security & Global Change, 23*(3), 315–334.

Whitehead, F. (2014). The first climate justice summit: A pie in the face for the Global North. *The Guardian.* http://www.theguardian.com/global-development-professionals-network/2014/apr/16/climate-change-justice-summit. Accessed 27 March 2016.

Wolinsky, Y. (1994). *International bargaining under the shadow of the electorate.* Ph.D. dissertation, Department of Political Science, The University of Chicago, Chicago, IL.

World Bank. (2013a). CO2 emissions per capita data. http://data.worldbank.org/indicator/EN.ATM.CO2E.PC. Accessed 10 August 2014.

World Bank. (2013b). Development data group, world development indicators 2012. http://data.worldbank.org/sites/default/files/wdi-2012-ebook.pdf. Accessed 10 August 2014.

Index

A
Academia, 6, 89, 90, 91

B
BASIC, 10, 11, 26, 27, 40, 48, 51, 59, 61–62, 64, 85, 87, 91, 100, 124
BRICS, 9, 24–25, 26–27, 40, 85

C
China, 9–10, 13, 24, 26, 36, 46–49, 51, 59, 61, 64, 85, 87, 91, 100, 117, 124–126
Civil society, 6
Coal, 31–35, 38, 53, 55, 58, 72–75, 89, 91, 94, 121–122, 124, 125
Common but differentiated responsibilities, 7, 51, 72, 101–102, 119
Conference of Parties, 8, 26, 47, 90, 100, 118
COP, *see* Conference of the Parties

Copenhagen, 8, 10–12, 26, 47, 100, 105, 107

D
Doha, 12, 47, 53–54, 55, 57, 100, 102
Dubash, 10, 11, 30, 31, 40

E
Electrification, 28–30
Energy security, 7–8, 27, 31–33, 37–38, 39, 41, 48–49, 50, 52, 53, 54–62, 72–76, 77, 86, 90, 101, 105, 107, 125

G
G-77, 22–24, 26, 39–40, 49, 52, 54–55, 59, 64, 81–84, 90–91, 100, 107
Gandhi, 7, 17, 22–23, 29, 93, 101, 103

I

Identity, 3–7, 10–12, 15, 16, 18, 20, 22, 26, 28, 30–31, 34, 36, 38–41, 46, 48–70, 72, 74, 76, 78, 80–82, 84, 85, 88, 90, 91, 94, 96–100, 102–108, 111–128

Indian Institute of Technology, 33

K

Kyoto, 12, 16, 26, 83, 103–105

L

Levinas, 3–5

M

Maslow, 5–6
Media analysis, 6, 47
Modi, 116–122

N

Natarajan, 8, 93, 102, 106, 123
National Action Plan, 106
National Mission on Solar Energy, 35
Nuclear, 31, 37, 75–76, 89, 92

O

OECD, 9, 28
Oil, 31, 37–38
Ontology, 3–13

P

Per capita emission, 7–8, 10, 21, 122

Prime Minister, 7–8, 17, 22, 29, 93, 103–104, 116
Private sector, 6, 30, 65, 75, 88–91

R

Ramesh, 8, 10, 34, 53, 87, 102–103, 106
Renewables, 31, 35–36, 39, 72–73, 75, 89, 91, 94, 126
Russia, 9–10, 24, 26, 85

S

Singh, 8, 33, 114
Solar, 35, 36, 73, 78, 106, 116, 121, 126

T

Technology transfer, 7, 17, 24, 89, 105, 106, 121–122

U

UNFCCC, 9, 11–12, 16, 26, 105–106, 119
United Nations, 7–9, 12, 16, 17, 22–23, 38–39, 93, 100, 104–106, 118
United States of America, 11–12

V

Voluntary targets, 106

W

Wind, 35, 73, 112, 121, 126

The manufacturer's authorised representative in the EU is Springer Nature Customer Service Centre GmbH, Europaplatz 3, 69115 Heidelberg, Germany. If you have any concerns regarding our products, please contact ProductSafety@springernature.com

Printed and bound by CPI Group (UK) Ltd, Croydon, CR0 4YY
23/03/2026
02076449-0006